51.00
100K

AIP CONFERENCE PROCEEDINGS 161

RITA G. LERNER
SERIES EDITOR

ELECTRON SCATTERING IN NUCLEAR AND PARTICLE SCIENCE

IN COMMEMORATION OF THE
35TH ANNIVERSARY OF THE
LYMAN-HANSON-SCOTT EXPERIMENT

URBANA, IL 1986

EDITORS:

C. N. PAPANICOLAS, L. S. CARDMAN &
R. A. EISENSTEIN
UNIVERSITY OF ILLINOIS AT
URBANA-CHAMPAIGN

AMERICAN INSTITUTE OF PHYSICS   NEW YORK 1987

Authorization to photocopy items for internal or personal use, beyond the free copying permitted under the 1978 US Copyright Law (see statement below), is granted by the American Institute of Physics for users registered with the Copyright Clearance Center (CCC) Transactional Reporting Service, provided that the base fee of $3.00 per copy is paid directly to CCC, 27 Congress St., Salem, MA 01970. For those organizations that have been granted a photocopy license by CCC, a separate system of payment has been arranged. The fee code for users of the Transactional Reporting Service is: 0094-243X/87 $3.00.

Copyright 1987 American Institute of Physics

Individual readers of this volume and non-profit libraries, acting for them, are permitted to make fair use of the material in it, such as copying an article for use in teaching or research. Permission is granted to quote from this volume in scientific work with the customary acknowledgment of the source. To reprint a figure, table or other excerpt requires the consent of one of the original authors and notification to AIP. Republication or systematic or multiple reproduction of any material in this volume is permitted only under license from AIP. Address inquiries to Series Editor, AIP Conference Proceedings, AIP, 335 E. 45th St., New York, NY 10017.

L.C. Catalog Card No. 87-72403
ISBN 0-88318-361-7
DOE CONF-8610239

Printed in the United States of America

Contents

Session A
Chair – D. G. Ravenhall

Welcoming Address 2
    R. A. Eisenstein
The Need for Accelerating Electrons 5
    D. W. Kerst
Beyond Mott Scattering 12
    A. O. Hanson
A Microscope for Nuclei and Nucleons 23
    R. Hofstadter

Session B
Chair – H. Feshbach

Electron Scattering and Few-Nucleon Systems 38
    B. Frois
Scaling—From Momentum Distributions to Bound Nucleon Properties 58
    I. Sick
Electron Scattering and Correlations in Nuclei 74
    V. R. Pandharipande
The Electron Probe at Very High Energies (*experiment*) (*paper submitted too late for inclusion in this volume*)
    J. G. Branson

Session C
Chair – C. deVries

Electron Accelerators for Research at the Frontiers of Nuclear Physics 100
    H. A. Grunder, B. K. Hartline, and S. T. Corneliussen
Electron Accelerators at the Frontiers of Particle Physics 123
    B. D. McDaniel
Nuclear Physics with Internal Targets in Electron Storage Rings 138
    R. J. Holt
Accuracy of the Electron Probe 153
    D. G. Ravenhall

Session D
Chair – D. R. Yennie

Higher Resolution and Higher Momentum Transfer: New Insight into Nuclear Structure 174
    J. H. Heisenberg
Intermediate Energy Nuclear Physics with Electrons 189
    E. J. Moniz
The Future of High Energy Electron Beam Physics (*theory*) 206
    J. D. Bjorken
Symposium—35 years of Electron Scattering (*concluding remarks*) 213
    J. D. Walecka

List of Participants 225

Proceedings of the Workshop on

ELECTRON SCATTERING IN NUCLEAR AND PARTICLE SCIENCE

Nuclear Physics Laboratory
and
Physics Department
University of Illinois at Urbana-Champaign
October 23–24, 1986

Preface

It was truly a unique event. For two days we had the opportunity to gather in the same room the pioneers, the experts, and the students of electron scattering. The exchange of data, ideas, experience, hopes, and aspirations was very intense. At the end we all felt it was well worth the effort.

The occasion was the 35th anniversary of the first electron scattering experiment to observe nucleonic structure. (Hanson, Lyman, and Scott, *Phys. Rev.* **84**, 525 [1951]). It is really remarkable that in such a highly developed field as electron scattering, it is still possible to gather together the inventor of the betatron (Kerst), pioneers such as Hanson and Hofstadter, and a number of today's graduate students, and engage with them in lively discussions not only about the history of the field but also about its future. It is a reflection both of the maturity of the field and of its youthful vigor.

The symposium occurred at a time of great promise for the field, with many large scale developments already underway. New, powerful and novel accelerating schemes are being employed to provide us with beams still more intense, precise, and energetic and of the highest duty factor; projects of huge scale such as LEP, SLC, and CEBAF, and more modest but highly innovative designs are expected to yield answers to many of the most intriguing questions of our field. The excitement of this bright future permeated the meeting; it coupled with the reminiscences and the achievements of the past to produce a unique blend of optimism. It is hoped that these proceedings provide a perspective on our accomplishments and on our future directions.

Many people have contributed their talent, time, and effort to see that this symposium was a success. I would like to thank especially Phil Krick, Evan Jones, and Penny Sigler for their central roles in dealing with every aspect of the organization of this event. Other personnel of the Nuclear Physics Laboratory and the Physics Department provided us with their experience and help, which proved indispensible during the days of the event.

Finally, thanks are due to the speakers, session chairmen, and participants, who made this symposium a reality.

C. N. Papanicolas
L. S. Cardman
R. A. Eisenstein

Scattering of 15.7-Mev Electrons by Nuclei*

E. M. LYMAN, A. O. HANSON, AND M. B. SCOTT†

Department of Physics, University of Illinois, Urbana, Illinois

(Received July 3, 1951)

Electrons removed from the 20-Mev betatron are focused to a 0.08-inch spot about 10 feet from the betatron by a magnetic lens. The electrons impinge on thin foils at the center of a highly evacuated scattering chamber having a diameter of 20 inches. Elastically scattered electrons, selected by a ⅜ inch×2 inch aperture, are focused by means of a 75° magnetic analyzer with 3 percent energy resolution and are detected by coincidence Geiger counters. Corrections are applied for multiple scattering and for energy losses which remove the electrons from the range of energies accepted by the detector arrangement. The scattering cross section for gold at 150° is found to be about 2.6 times that given by Mott's formula in the Born approximation and about one-half of that expected for the scattering by a point nucleus. This result is in good agreement with the calculations for electrons of this energy if the nuclear charge is assumed to be distributed uniformly throughout the nuclear volume.

The results for the scattering from C, Al, Cu, and Ag are also in agreement with the assumption of a uniformly distributed nuclear charge within the uncertainties involved in the theory and the experimental results.

INTRODUCTION

THE scattering of fast electrons by nuclei has been the subject of a large number of researches. The previous work in the energy range above 10 Mev has been done primarily with cloud chambers. The results are quite divergent but could be considered to be in qualitative agreement with theoretical calculations.[1]

At lower energies some accurate measurements have been made using electrons accelerated by an electrostatic generator to energies up to 2.25 Mev. These results, reported by Van de Graaff, Buechner, and Feshbach, for several elements and for a number of angles up to 50° are in good agreement with calculations.[2] The more recent work of Champion and Roy and that of Sigrist[3] indicate that the scattering at larger angles is also in agreement with calculations, although others report divergent results.[4]

It appears that the remaining discrepancies at these energies are due to experimental difficulties, and it will be assumed in this work that the scattering of electrons having energies up to 3 Mev are described by complete calculations[5,6] based on Mott's formulas for the scattering by a point charge.

At sufficiently high energies, Rose, and more recently Elton, have shown that the scattering of electrons by nuclei would be considerably modified by the fact that the size of the nuclear charge distribution is no longer small compared to the electron wavelength.[7]

An experimental investigation of the scattering of high energy electrons was made feasible by the successful extraction of the electron beam from the 20-Mev betatron. Preliminary results have been reported briefly[8] and have been compared with the accurate calculations by Elton.[9]

I. APPARATUS AND EXPERIMENTAL METHOD

Production, Extraction, and Focusing of the Electron Beam

The experimental arrangement is shown schematically in Fig. 1. When electrons accelerated in the betatron donut reach 15.7 Mev, as determined by a flux

FIG. 1. Schematic showing betatron and scattering chamber.

---

* Supported by the joint program of ONR and AEC.
† Now at Massachusetts Institute of Technology, Cambridge 39, Massachusetts.
[1] Randels, Chao, and Crane, Phys. Rev. **68**, 64 (1945). See also reference 18, p. 83.
[2] Van de Graaff, Buechner, and Feshbach, Phys. Rev. **69**, 452 (1946); Buechner, Van de Graaff, Sperduto, Burrill, and Feshbach, Phys. Rev. **72**, 678 (1947).
[3] F. C. Champion and R. R. Roy, Proc. Phys. Soc. (London) **61**, 532 (1948); W. Sigrist, Helv. Phys. Acta **16**, 471 (1943).

[4] Alichanian, Alichanow, and Weissenberg, J. Phys. (USSR) **9**, 280 (1945); W. Bothe, Z. Naturforsch. **4a**, 88 (1949).
Note added in proof:—Paul and Reich have recently reported general agreement with theory at 2.2 Mev except for somewhat low values for gold at 90° and 120°. (Private communication. Work to appear in *Zeitschrift für Physik*.)
[5] J. H. Bartlett and R. E. Watson, Proc. Am. Acad. Arts Sci. **74**, 53 (1940).
[6] W. A. McKinley and H. Feshbach, Phys. Rev. **74**, 1759 (1948).
[7] M. E. Rose, Phys. Rev. **73**, 279 (1948).
[8] Lyman, Hanson, and Scott, Phys. Rev. **79**, 228 (1950); Phys. Rev. **81**, 309 (1951).
[9] L. R. B. Elton, Phys. Rev. **79**, 412 (1950); Proc. Phys. Soc. (London) **A63**, 1115 (1950).

Sponsors

U. S. National Science Foundation
Physics Department, University of Illinois
Engineering College, University of Illinois
Center for Advanced Study, University of Illinois
Research Board, University of Illinois

Organizing Committee

L. S. Cardman
R. A. Eisenstein
V. R. Pandharipande
C. N. Papanicolas (Chair)

SESSION A

WELCOMING ADDRESS

Robert A. Eisenstein
University of Illinois, Urbana, Il. 61801

It is indeed a pleasure to welcome all of you here on behalf of the Nuclear Physics Laboratory. We hope that your few days here will be most pleasant ones, and that during your brief stay you will become better acquainted with our activities and our plans for the future.

The reason for the Symposium is to mark in a small way the astonishing role that developments in electron scattering experiment and theory have played in nuclear and particle physics during the last 35 years. We are here also to celebrate the success of some of the founders of our field, Al Hanson and Ernie Lyman, whose ideas have had far-reaching influence both at the NPL and outside of Illinois. We are also pleased to welcome Donald Kerst, now at the University of Wisconsin, who developed the betatron at Illinois in the early '40s. With pleasure we also welcome Bob Hofstadter, whose pioneering experiments at Stanford gave rise to the great Stanford tradition in our field, and one of the great success stories of physics in the latter part of the twentieth century.

So, the Symposium is not really a conference in the usual sense of that word. Instead, on the one hand we are celebrating the history of our field, and on the other we are providing brief glimpses of both the most important issues in modern nuclear and particle physics as well as the new developments in accelerator technology that makes it possible to study them. This morning we will begin with Lord Rutherford and the origins of electron scattering in nuclear and particle physics; this afternoon we will conclude with a discussion of one of the largest physics experiments ever mounted, the L3 detector at LEP. Both today and tomorrow we will discuss the physics and the technology, emphasizing essential link between the two.

In many countries the world over, nuclear and particle physicists have placed much of their scientific future in the development of new, large scale facilities to do electromagnetic physics research on the structure of nuclei and particles. Of course, this is because real and virtual photons are the quanta of our best understood theory, quantum electrodynamics. Research with these tools has provided an incredibly rich, precise, and reliable view of nuclear and particle structure. We count among the many successes of the field the unraveling of the essential aspects of nuclear charge and current densities, the elucidation of the nature of the nucleon's structure, including its composition as a cluster of partons, essential contributions to the discovery of the J/ψ and the spectroscopy of the Υ, tests of the Standard Model, and the wonderful confusion of the EMC effect! There is every indication that such interesting developments will continue to be a part of the future.

Over the years, the experimental side of our science has advanced dramatically because of essential, even revolutionary, breakthroughs in accelerator physics and related instrumentation that has allowed great progress toward increasing the energy, intensity and duty-factor of these instruments. Detectors capable of dealing with such large energies, luminosities, and multiplicities have of course become indispensible parts of these facilities and now often rival or exceed the complexity and cost of entire laboratories of an earlier time.

Today we are in the process of putting into place in the US and world-wide a structured program for further progess in electromagnetic nuclear and particle science. This program will explore in detail the strong interaction of quarks and gluons from the region of asymptotic freedom to the region of complete confinement. In the high energy domain, at CERN, the Large Electron Positron Collider ring will bring particle beams of 50-60 GeV on 50-60 GeV into collision at four intersection regions for further research into the nature of the Z_0 particles, searches for the top quark and the Higgs boson, and other tests of the Standard Model. Operation is scheduled to begin in 1989. A similar program is envisioned at SLAC, where the single pass collider (SLC), operating at energies of 50 GeV on 50 GeV will hopefully begin operation in 1987. At Cornell, CESR will continue its beautiful studies of the spectroscopy of the Υ and b-meson regions with the accelerator upgraded to high luminosity beams of up to 8 GeV on 8 GeV and the new CLEO detector on line. In Hamburg, DESY will commission the HERA project in 1990. At SLAC, the SPEAR ring will continue to be used for investigation of physics in the J/ψ and $b\bar{b}$ regions with Mark III, and PEP will operate in its collider mode at COM energies up to 29 GeV.

In the domain of nuclear and intermediate energy physics, the structured US plan for electromagnetic nuclear physics research has one major goal: to make possible for the first time investigation of fundamental problems in nuclear physics that were heretofore simply unaccessible because of energy, intensity, or duty factor limitations. It is already clear, based on limited work done at existing accelerators, that the proposed new facilities will make available a very broad new front along which to attack these problems. These investigations, which should provide important additional clues to the nature of the strong interaction many-body problem, include studies of both discrete states and giant resonance phenomena, studies of the nature of deep-lying hole states in nuclei, further investigation of the nature of NN correlations in nuclear matter, a detailed study of meson degrees of freedom in nuclei, spin studies of nuclei using polarized electrons and photons, and searches for explicit signatures of quarks in nuclei. The high duty factor aspect of the new facilities opens up a completely new region of physics to be explored, making possible coincidence experiments that will enable us to decode the wavefunction of the nuclear state in ways that were simply impossible before. I believe that "renaissance" is not too dramatic a word to describe these possibilies.

The new generation of electron accelerators for nuclear physics is typified in this country by the CEBAF, MIT-Bates, and Illinois projects, which cover three rather distinct areas of nuclear physics interest. At present, CEBAF, which is about to begin construction, will provide a primary beam energy of 4 GeV, 240μA current, and a 100% duty factor using two superconducting linac accelerator sections and 4 recirculations. MIT-Bates proposes to use a recirculator and storage ring technology to produce beams up to 1 GeV in energy with 100 μA current and 80% duty factor. Their plans also call for an internal gas jet target facility, and polarized electron beams. Illinois will build a two-stage cascade microtron with room-temperature accelerating linacs to provide beams of 450 MeV, 100 A current and 100% duty factor. Multiple simultaneous beams will allow for extremely efficient use of this facility.

Abroad, facilities with similar objectives, such as those at Bonn, Darmstadt, Frascati, Lund, Mainz, Moscow, Saskatoon and Tohoku, will commence operation soon. Other laboratories, such as NIKHEF and Saclay, are also planning construction projects aimed at exploiting these exciting new areas. It is the beginning of a new era in our field.

Many of these new and exciting new initiatives will be discussed today and tomorrow. I will, instead of continuing to preach to the converted, take my seat and enjoy what all of our speakers and participants have to say. Thank all of you for coming!

THE NEED FOR ACCELERATING ELECTRONS

Donald W. Kerst
University of Wisconsin, Madison, WI., 53706

ABSTRACT

Photons for nuclear disintegration experiments were not abundantly available in the early days of nuclear physics, whereas accelerated ions led the way. When electrons could be accelerated into the 20-30 MeV range, they found application not only to nuclear disintegration of the elements of the periodic table but also to x-ray radiography and to deep therapy. Energies of interest for probing nuclear structure by electron scattering and for meson production followed soon after. The elementary nature of the electron has now made it a valuable tool for present day particle physics; and the synchrotron radiation, which is an obstacle for some accelerating processes, has become a much sought after source of photons for experiments at atomic structure energies.

In the 1930's, when experiments on nuclear disintegration became possible, the experiment of Chadwick and Goldhaber on the photo-disintegration of deuterium by thorium C" gamma rays was important news.[1] This brought a new use for gamma rays in nuclear research into being. Deuterium was a new isotope at that time; so it was not readily available. This particular experiment was done when we were scattering protons by protons at Wisconsin; but we were so intent with our efforts to accumulate proton-proton scattering data that we could not explore many of the other experimental possibilities which were developing at the time. However, we had a lot of experience with p-gamma production, and we had also in our Chemistry Department a large heavy water project. Professor Hall of Chemistry had set up a large bank of electrolytic cells just outside the Physics Department, and he was using our power supply to concentrate heavy water electrolytically; so we had the 17 MeV gamma rays from lithium bombarded by protons, and we had an available source of deuterium -the necessities for looking at the photo-disintegration of deuterum. There was a strong temptation to see if an experiment could be bootlegged into our busy experimental schedule; so with a very low priority, an ionization chamber system was constructed into which a gamma ray target of gelatin containing heavy water was to be installed. The chemists had given us enough water to make films. In the brief opportunities we had to test with heavy water gelatin we never saw the background count down where we wanted it. There were always large background pulses; so the accumulation of scattering data continued to take all our time. No doubt we had let an exciting series of experiments slip by, as we realized when we read Bothe's and Gentner's publication[2] about the photo-disintegration of numerous other elements which could have caused the background in our chamber.

Usually experimenters with accelerators were using positive ions for the production of nuclear disintegration; but, at Notre Dame, Collins and Waldman devoted their Van de Graaff electrostatic accelerator to experiments with electrons and their x-rays. They had almost two million volt electrons, and among the observations they made were the electro-disintegration and the photo-disintegration of beryllium and the excitation of indium. Electrons of sufficient energy to produce other nuclear effects were not available from accelerators. The experiments were dangerous, because, with an electron beam from an accelerator, x-rays can be sufficiently intense to damage an unprotected experimenter, and the electrons have a long enough range to scatter over walls and around the vicinity before they are stopped in the air.

These early nuclear experiments with electrons had the attention of experimenters; but, at the same time, there were other uses of electrons of practical importance which were very eveident. There were radiographic possibilities for these accelerators when used to generate x-ray beams, since industrial x-ray machines only went up to 400 kilovolts while the absorption coefficient in iron has a minimum above 20 million volts. Van de Graaff turned his attention mainly to developing the electrostatic accelerator in the two million volt range for this industrial purpose. High voltage for the medical application of x-rays for cancer therapy was another possible application for electrons.

Thus, for several purposes, more energetic electrons were desirable. For the

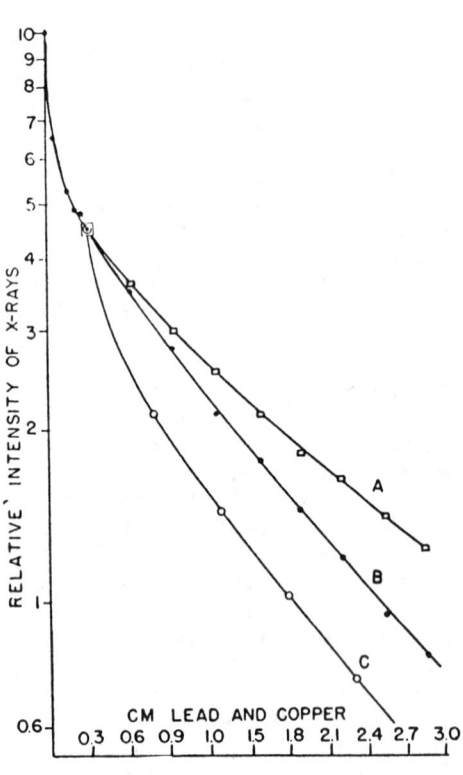

Fig. 1. One of the first uses of the Betatron was the measurement of absorption curves for copper (A) and lead (B). C is the lead absorption curve obtained earlier from the MIT electrostatic machine. This figure is taken from ref. 10 [editors].

penetration of iron, one could use ten times the energy of our early electrostatic accelerators so that the x-ray penetration would be the optimum for iron. For medical physics, two million volt electrons would just penetrate a centimeter into tissue; so one wanted an order of magnitude more range if he used the accelerated electron beam directly for treatment or if the x-rays produced were to be used in deep tissue.

For application to nuclear problems there were just a few cases which could be reached by the energy of the x-rays of these early electrostatic machines--deuterium and beryllium, for example. Otherwise, for the nuclear problem we had to use gamma rays from nuclear disintegration, such as the 17 million volt photons from protons on lithium.

It was frustrating to have available for nuclear experiments ions from cyclotrons and from electrostatic machines and to have neutrons which could produce nuclear disintegrations directly but to have only low intensity secondary photons from nuclear disintegrations or intense electron accelerator x-rays but of only 2 MeV and with a continuous bremsstrahlung spectrum.

It was clear that if we had energies for electrons and their resulting bremsstrahlung in the 20 MeV range there would be enough energy for photo-disintegration of the elements of the periodic table and later with somewhat more energy giant resonances could be excited.[4]

At the University of Illinois we were considering expanding the nuclear facilities with a larger cyclotron, and at the same time we were wondering what could be done to accelerate electrons by magnetic induction. Most universities were adopting the cyclotron for providing high speed ions approaching 20 MeV for nuclear experiments. The first model induction accelerator, for the betatron, was not much good for any application, but it made clear how to proceed to larger energies sufficient for nuclear physics and, additionally, for the other practical applications.

During the first trials when the betatron began to work, we learned important lessons in accelerator lore, while trying to establish its energy by detecting the nuclear disintegration of beryllium. If we had two million volts in electron energy, then a large chunk of beryllium in the x-ray beam should give us photo-neutrons and thus establish that we were above the 1.7 MeV threshold for photo-disintegration. The beryllium would be put in place, and the whole accelerator covered by paraffin blocks, with a slow neutron counter to detect the neutron production, but no neutrons or x-rays could be produced in this configuration. The whole assembly was torn down and re-built several times, but we couldn't produce x-rays. This was when we discovered that a block of beryllium metal placed close to the source of x-rays produces sufficiently large field error by its eddy currents that it destroyed the orbits and prevented acceleration of the electrons. It was a very important early lesson on the sensitivity of an orbit to field errors. The intensity from this little machine was not very great. It was equivalent to about two grams of radium placed at the target.

BETATRON

ORIGINAL AT ILLINOIS
OPERATED JULY 15, 1940

HIGH-ENERGY ELECTRONS ARE USEFUL BOTH AS FREE ELECTRONS AND AS A SOURCE OF EQUALLY HIGH-ENERGY X-RAYS. FOR YEARS SCIENTISTS TRIED TO BUILD A MACHINE TO ACCELERATE ELECTRONS TO HIGH ENERGIES BY USE OF A MAGNETIC FIELD. PROF. DONALD W. KERST SUCCEEDED AT THE UNIVERSITY OF ILLINOIS, JULY 15, 1940. HIS FIRST INSTRUMENT PRODUCED 2,300,000-VOLT X-RAYS.

HIS SECOND MACHINE, COMPLETED IN 1941, PRODUCES 22,000,000-VOLTS. IT IS THE PROTOTYPE OF THE COMMERCIAL BETATRON DEVELOPED DURING THE WAR AS AN X-RAY MACHINE FOR INDUSTRY. IT IS USED SCIENTIFICALLY FOR RESEARCH IN NUCLEAR PHYSICS, AND MEDICALLY FOR ATTACKING CANCER. IT CAN PRODUCE EITHER X-RAYS OR AN ELECTRON BEAM. AN IMPORTANT FEATURE IS PRECISE CONTROL OF THE ENERGY PRODUCED.

A 300,000,000-VOLT BETATRON NOW IS BEING BUILT AT ILLINOIS TO PRODUCE COSMIC RAYS IN THE LABORATORY AND OPEN ENTIRELY NEW DOORS TO SCIENCE. THE NAME "BETATRON" COMES FROM THE SYMBOL "BETA" USED BY SCIENTISTS TO INDICATE HIGH-SPEED ELECTRONS AND "TRON" MEANING "A DEVICE FOR."

UNIVERSITY OF ILLINOIS

An early flier explaining the operation and possible uses of betatrons.

But a copy of this early betatron driven up to 4.5 MeV was flown to the Woolwich Arsenal in England by H. W. Koch during the war for radiography, where it subsequently was operated as the first synchrotron by Goward and Barnes in mid-1946.[5]

In these present times, when efforts are being made to raise the current in betatrons by orders of magnitude, questions are asked about whether early experiments were done to raise the intensity by addition of toroidal field and whether space charge neutralization was tried,--the answer to both questions is "yes". In the first weeks of the life of the betatron, both experiments were done without encouraging results. The most rewarding efforts were other improvements of injection; so that was the effort pursued. We know now that both of these methods can be employed in accelerators with very intense beams, but the explorations have just begun.

Further, early lessons about field errors concerned electrostatic perturbations. An electrostatic perturbation of 300 β volts per centimeter is equivalent to a one gauss field error; so an insulated spot on the interior of the accelerating chamber charged up by the injected beam could cause a force equivalent to many gauss and could destroy the beam.

When the range of photon energies passed 20 MeV, the several applications mentioned were all possible,--radiography, therapy, and nuclear photo-disintegrations. The source was not especially convenient for nuclear physics although one could adjust the peak energy of the photon spectrum very precisely. Nuclear threshold determinations were limited because of the continuous nature of the x-ray spectrum until the development of the elegant monochromator method by which the radiation straggled electron was observed in coincidence with the x-ray result to establish the photo-energy precisely. This method was first used at Cornell[6] and used extensively at Illinois.[7]

The application to radiography was made by the Illinois group during the war for Los Alamos and for government arsenals. The porcelain accelerating chamber was developed and permanently sealed off to make this application practical. Likewise the group developed the techniques of deep radiation (22 MeV) therapy and treated the first patient in the Physics Laboratory in collaboration with a local hospital (Carle Foundation).[8]

Soon it was natural to consider the possibility of producing the meson. When this question came up, we knew it would take approximately 100 MeV for the meson. The π meson had not yet been recognized. It was unfortunate that that figure stuck at the time the 100 MeV betatron was made at General Electric. We should not have been so optimistic in our design at that time, but, when thinking of that threshold, 100 MeV seemed like infinity, and we knew that the centripetal radiation acceleration, which we now call synchrotron radiation, would not be too bad at that energy.

In the period of the betatrons at Illinois where nuclear physics was being done with electrons and photons, our colleagues elsewhere who used ions instead of electrons for disintegration experiments were always envied. Electron accelerators did not

seem to be as fashionable as ion accelerators. There was more
company for the electron community when the meson producing
electron synchrotrons came along after the war; but I'm sure we
never imagined that the electron deserved the high status of being
so elementary and as interesting as today. The electron
scattering experiments[9] which we are celebrating at our current
symposium were exploiting this character of the electron. So we
now have present day interest in electrons for their very
elementary nature, heroic efforts to devise ways to exploit
electron energies and intensities well beyond what we now have,
and, in the course of these developments, another community of
physicists--taking advantage of the disadvantage of electrons--has
grown up around the accelerators using synchrotron radiation for
light sources. They are asking for dedicated and conveniently
scaled sharp sources of light for solid state, biological, and
industrial application.

The author in front of the 22 MeV betatron examining the original
2.3 MeV accelerator [editors].

References

1. J. Chadwick and M. Goldhaber, Proc. Roy. Soc. (London) 151A, 479 (1935).
2. W. Bothe and W. Gentner, Z. Physik, 106, 236 (1937).
3. G. B. Collins, B. Waldman, and E. Guth, Phys. Rev. 56, 876 (1939).
4. M. Goldhaber and E. Teller, Phys. Rev., 74, 1046 (1948).
5. F. K. Goward and D. E. Barnes, Nature, September 21, 1946, p. 413.
6. J. W. Weil and B. D. McDaniel, Phys. Rev. 92, 391 (1953).
7. J. S. O'Connell, P. A. Tipler, and P. Axel, Phys. Rev. 126, 228 (1962), and references therein.
8. H. Quastler, G. D. Adams, G. M. Almy, S. M. Dancoff, A. O. Hanson, D. W.Kerst, H. W. Koch, L. H. Lanzl, J. S. Laughlin, D. E. Riesen, C. S. Robinson, Jr., V. T. Austin, T. G. Kerley, E. F. Lanzl, G. Y. McClure, E. A. Thompson, L. S. Skaggs, Amer. Jour. Roentgenology and Radium Therapy 61, 591 (May, 1949).
9. E. M. Lyman, A. O. Hanson, and M. B. Scott, Phys. Rev. 84, 626 (1951).
10. D. W. Kerst, Phys. Rev. 60, 47 (1941).

BEYOND MOTT SCATTERING

A. O. Hanson
University of Illinois, Champaign, IL 61820

I am pleased to be here as one of the co-workers on the Illinois experiment being commemorated. I am also pleased that Ernie Lyman and Merrill Scott, who did most of the work on this experiment, are with us on this occasion. Lester Skaggs and Larry Lanzl who worked with Don Kerst on the extraction of the beam are also here, as are Jim Leiss, Fred Mills, and Ned Goldwasser who worked with us on this or other experiments with the 15.7 MeV electron beam.

Although the title suggested for this talk "Beyond Mott Scattering" seems to allow me to speculate into the future, I discovered that in 1951 we were already well beyond the year of 1929 when Mott published the well known formula for the scattering of relativistic electrons by nuclei. I have therefore chosen to examine some of the work before Mott and leading up to the review of the scattering work at Illinois.

In looking way back I was pleased to find a statement by Ernest Rutherford in the introduction to his classic paper of 1911 which is still relevant.[1]

> "Since the α and β particles traverse the atom, it should be possible from a close study of the nature of the deflexion to form some idea of the constitution of the atom to produce the effects observed. <u>In fact, the scattering of high-speed charged particles by the atoms of matter is one of the most promising methods of attack of this problem.</u> The development of the scintillation method of counting single α particles affords unusual advantages of investigation, and the researches of H. Geiger by this method have already added much to our knowledge of the scattering of α rays by matter."

He derived his well known formula for the scattering of charged particles by a fixed central Coulomb field. Namely

$$\sigma_R(\theta) = \left[\frac{Zze^2}{2mv^2}\right]^2 \csc^4 \theta/2.$$

Here one can see that the angular distribution is independent of the central charge Z and the energy of the incoming particle. By 1913 Geiger and Marsden had measured the angular distribution over angles from 7.5° to 150° using the simple apparatus shown as Fig. 1. It consists of a collimated alpha source, a thin gold foil and a scintillation detector observed through a microscope. They found good agreement between the observed and calculated angular distributions over the entire range of angles which indicated that the radius of the gold nucleus was smaller than the distance of closest approach, about 5×10^{-12} cm, and established the nuclear model of the atom. One might note here that one of the unexpected things in Rutherford's paper was that he didn't know the charge on the gold nucleus. He obtained a value of 97 to 114 from the preliminary scattering data and speculated that the nucleus might be made up of 49 alpha particles. We note that an alpha particle would need an energy of 26 MeV to reach the surface of the gold nucleus.

Rutherford also considered some of the observations on the scattering of electrons. These observations were rather qualitative but he concluded that they should be scattered according to the same formula. Progress in the study of the scattering of electrons was limited by the lack of monoenergetic electron sources as well as the lack of suitable discriminating detectors.

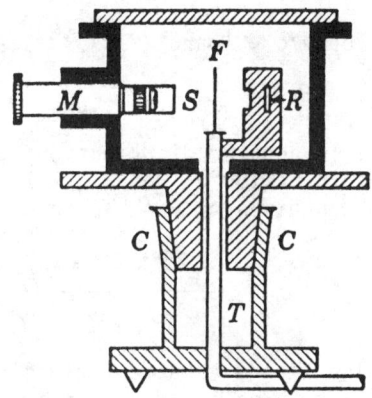

Fig. 1. The apparatus used by Geiger and Marsden for studies of alpha-particle scattering by nuclei. (1913)

The development of the quantum theory of the electron by Dirac in 1928 created a new interest in electron scattering. Mott was with Bohr in Copenhagen at that time and was concerned with demonstrating a way in which the spin of a free electron could be observed. In his recent biography he reproduces a letter home in which he describes the great pleasure in his contacts with Bohr. He ends the letter with:

"And he has his students come in the evening to talk and then walks them home, telling how he discovered his theory of spectra. And then its 1 a.m. perhaps. But it is funny that the spin of the electron can never be observed isn't it. Perhaps spin is only an illusion."

In his paper of 1929 Mott, however, did develop a formula for the polarization of electrons by Coulomb scattering as well one for the scattering of an unpolarized beam.[2] The formulas as he wrote them were:

$$|f|^2 + |g|^2 = \frac{Z^2 \varepsilon^4}{4m^2 v^4} (1 - \frac{v^2}{c^2}) [\operatorname{cosec}^4 \frac{\theta}{2} - \frac{v^2}{c^2} \operatorname{cosec}^2 \frac{\theta}{2}$$

$$+ d\frac{v}{c} \pi \frac{2\pi Z \varepsilon^2}{hc} \frac{\cos^2 \frac{\theta}{2}}{\sin^3 \frac{\theta}{2}} + \text{terms of order } \alpha^2],$$

and

$$fg^* - f^*g = \frac{Z^2 \varepsilon^4}{4m^2 v^4} (1 - \frac{v^2}{c^2})^{3/2} \frac{v}{c} 4\alpha \operatorname{cosec} \theta \log \operatorname{cosec} \frac{\theta}{2}.$$

The first equation with only the first two angular terms is referred as $\sigma_{Mott}(\theta)$. It approaches $\sigma_{Rutherford}$ for $v/c \ll 1$.

For v/c ≈ 1 and Z/137 << 1 σ_{Mott} can be written [3]

$$\sigma_{Mott} \approx \left(\frac{Ze^2}{2E_t}\right)^2 \frac{\cos^2 \theta/2}{\sin^4 \theta/2}.$$

This angular distribution agrees with σ_R at low angle but goes to zero at 180°. For v/c ≈ 1 and low Z a corrected approximation has been given by Feshbach and others[4]

$$\sigma_F = \left(\frac{Ze^2}{2E_t}\right)^2 \frac{\cos^2 \theta/2}{\sin^4 \theta/2} \left[1 + \frac{\pi Z}{137} \frac{\sin \theta/2 \,[1 - \sin \theta/2]}{\cos^2 \theta/2}\right].$$

The second equation determines the polarization of the scattered beam. The polarization increases with Z^2 and has a maximum around v/c = .6 for scattering by 90°.

One of the earliest attempts to check Mott's formula for the scattering of electrons was made by H. V. Neher at the California Institute of Technology in 1931.[5] He used an electron gun to supply a narrow beam of electrons at voltages up to 145 kV (β=.63). His experimental arrangement is shown as Fig. 2, which

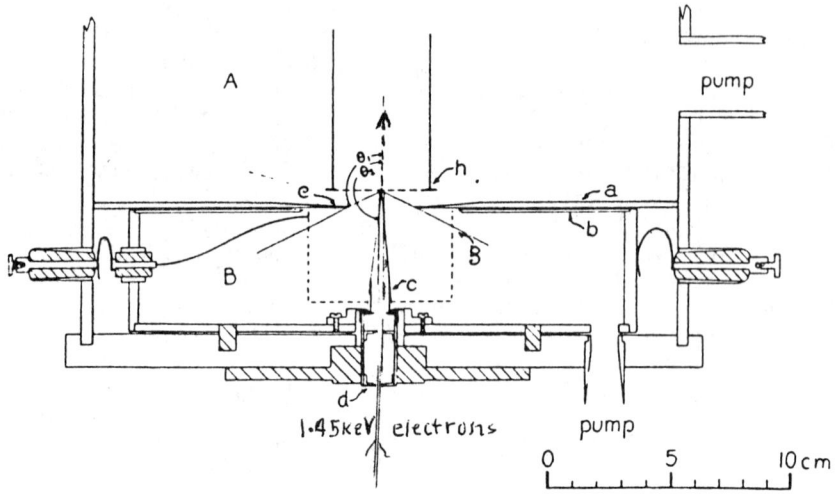

Fig. 2. Scattering chamber used by H. V. Neher at Cal Tech. in 1931 for measuring the back scattering of 145 keV electrons from aluminum, silver and gold through 95° to 173°. Note the scattering foil (h), the grid for repelling slow electrons (g) and the collector (b).

includes a charged wire grid at the entrance of the detector to repel low energy electrons. He observed the scattering from aluminum, silver and gold at backward angles between 95° amd 173°. He found that the angular dependence agreed better with Rutherfords equation. He made the interesting observation that secondary electrons coming from the foils were distributed according to the simple cosine law.

A number of people made measurements of electron scattering in cloud chambers and obtained results with very poor statistics in general agreement with calculations. Barber and Champion, in 1938 reported on the scattering on electron scattering from mercury from pictures of scattering events in cloud chamber containing a vapor of mercury dimethyl.[6] This work apparently stimulated Bartlett and Watson at Illinois in 1940 to sum Mott's exact expression for the scattering electrons for mercury (Z=80).[7] These calculations are also approximately true for gold (Z=79). Their calculated values are shown in Fig. 3. It can be

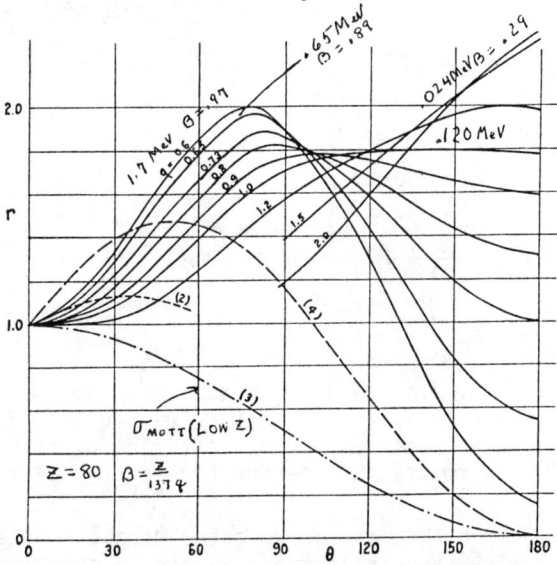

Fig. 3. Ratio of scattering calculated for Z= 80 by Bartlett and Watson to Rutherford scattering. Broken lines represent different approximations. Curve (3) represents the simple Mott formula for low Z and $\beta = 1$. (1940)

Table I. Comparison of the angular distribution calculated for Z = 80 by Bartlett and Watson and the experimental angular distributions observed in a cloud chamber by Barber and Champion for electrons of about 1 MeV. (1938)

Angular range	20°-30°	30°-60°	60°-180°
Experimental no. observed	37	60	8
Calculated no. normalized	37	27	9

seen that the scattering by high Z elements is much enhanced as compared with the simple Mott formula. The calculated angular distribution agreed with that observed by Barber and Champion. The relative numbers in the angular ranges 20°-30°, 30°-60° and 60°-180° were 37:27:9 as compared to the experimental ratios of 37:30:8. The absolute number of events, however, was much below the calculated number.

Mott[8] (and also Bartlett and Watson) evaluated the expression for the polarization of electrons scattered by high Z nuclei and for the asymmetry expected after double scattering by 90°. Although there were a number of attempts to observe this effect in cloud chambers it was not clearly demonstrated until 1942 by the double scattering experiment of Shull, Chase and Myers.[9] They used a 400 keV electron beam from a Van de Graaff electrostatic generator, scattered successively by 90° by thin gold foils. Their scattering arrangement is shown in Fig. 4. They found a

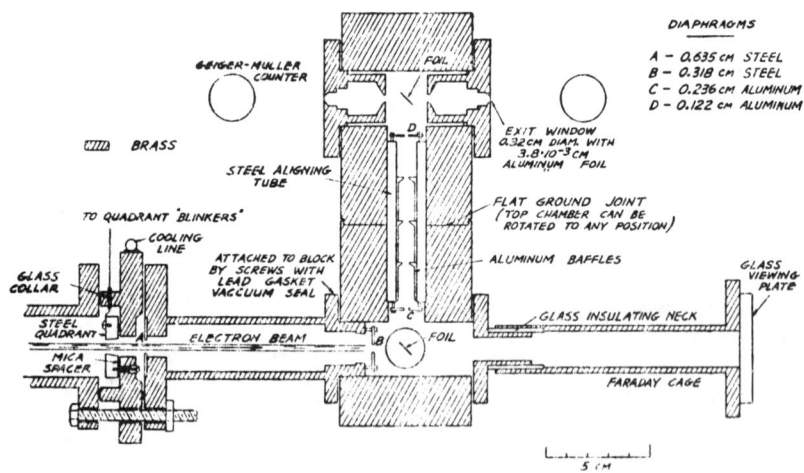

Fig. 4. The experimental arrangement of Shull, Chase and Myers for measuring the asymmetry due to polarization of 400 keV electrons by double scattering by gold foils. The right-left asymmetry is detected by Geiger-Muller counters. (1942)

right left asymmetry after double scattering of 12% in agreement with the Bartlett and Watson calculation. This point together with calculated curves is shown in Fig. 5. I may note here, that after the discovery of parity violation in 1956, the Mott asymmetry due to single scattering was useful in determining the polarization of beta decay electrons from various radioactive nuclides. At Illinois Frauenfelder et al., after transforming the longitudinal polarization of electrons from Co^{60} into a transverse one, found a scattering asymmetry of 35%.[10]

In 1946 an experiment was reported by Van de Graaff, Buechner and Feshbach on the scattering of electron beams with energies from 1.27 to 2.27 MeV.[11] Scattering was observed from aluminum, copper, silver and gold foils at angles of 20°, 30°, 40°, and 50°. Their scattering arrangement with the movable ionization chamber detector is shown in Fig. 6. The ratios of the experimentally measured cross sections to those calculated were in good agreement to within a few percent, and represented a significant check on the exact calculations for the heavier elements. Their data is displayed as a function of the distance of closest approach in units of e^2/mc^2 = 2.8 x 10^{-13} cm in Fig. 7. The radius of gold is about 3 units on this graph. Classically at this energy the electrons remain well outside of the nuclei and

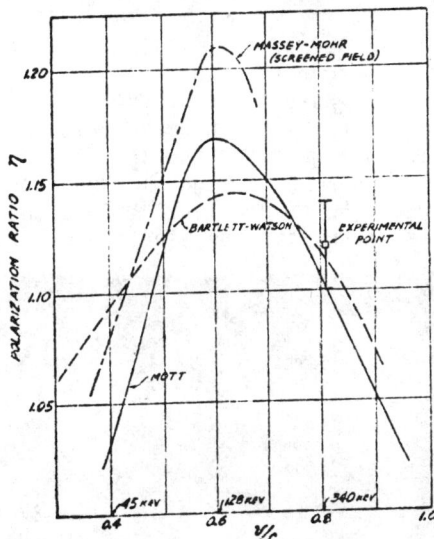

Fig. 5. Theoretical values for the polarization asymmetry as a function of v/c. The experimental point is from Shull, Chase and Myers.

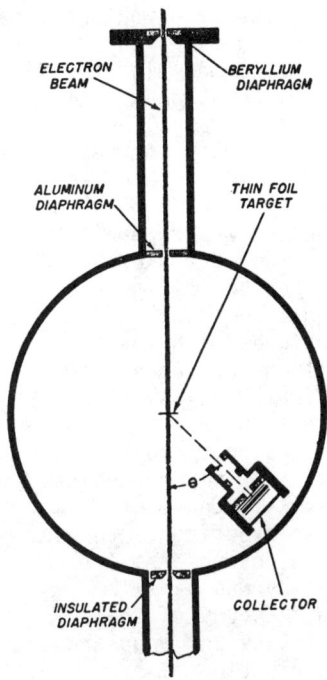

Fig. 6. Scattering chamber of Van de Graaff, Buechner and Feshbach. (1946)

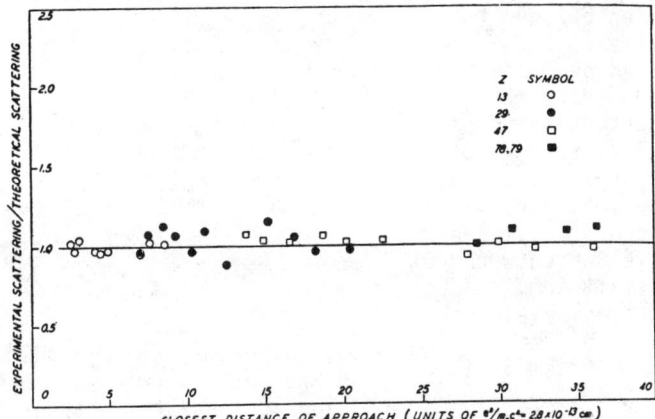

Fig. 7. Ratio of experimental to theoretical cross sections as a function of the distance of closest approach in units of $e^2/mc^2 = 2.8 \times 10^{-13}$ cm.

the scattering would not be sensitive to the sizes of the nuclei.

1946 was also the year when I arrived here at Illinois. The 20 MeV betatron was already a reliable source of 20 MeV bremsstrahlung. Some of the staff were busy studying the radioactivity induced in various elements by the bremsstrahlung. Others were interested in extracting the beam for use in cancer therapy. The removal of the electron beam by a magnetic shunt or "peeler" was reported by Skaggs, Almy, Kerst and Lanzl before the year was out. Kerst and Skaggs are shown with the electron donut and peeler in Fig. 8. Skaggs and Lanzl continued with very influential careers in Medical Physics. Lanzl is now the President of the International Organization for Medical Physics. One of the first experiments to use the electron beam for physics was Lanzl's thesis study of the intensity and angular distribution of bremsstrahlung from a number of materials.

Fig. 8. Don Kerst and Lester Skaggs with the peeler and electron donut.

About this time when we were thinking about other ways of using the electron beam, we were encouraged to consider electron scattering by nuclei by Rose and others who called attention to the possibility that at our energies the scattering might be considerably modified by the size of the nuclei which were no longer small compared to the electron wavelength ($R_{Au} \approx 9 \times 10^{-13}$ cm, $\lambda \approx 12 \times 10^{-13}$ cm at 16 MeV).

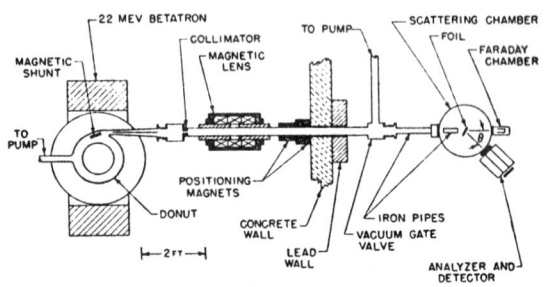

Fig. 9. Experimental arrangement of Lyman, Hanson and Scott showing the betatron, the focussing lens and the scattering chamber.

Lyman joined us at this point and was much involved in the design and construction of the scattering chamber, the associated magnetic analyzer and the coincident counter detector system.[12] The experimental arrangement for 30° scattering is shown in Fig. 9. A photograph of this arrangement

Fig. 10. A photograph of the laboratory showing the betatron and the equipment associated with the nuclear scattering experiment.

is shown as Fig. 10. One might note the massive cylindrical lens which focussed the beam on targets at the center of the scattering chamber. It was mounted in a water cooled oil bath since it took a lot of power. If I remember correctly it was operated by a rather large motor generator set capable of supplying 100 amperes at 100 volts. The lens was difficult to keep cool. It eventually burned up and was replaced by a modest quadrupole doublet.

A drawing of the scattering chamber and detector arrangement is shown in Fig. 11. The scattered beam from a point at the center of the chamber was selected by a well defined aperture in the chamber wall and was refocussed by the magnetic analyzer to a short line on the Geiger counters in coincidence. The counters accepted electrons with energy losses up to 3 percent. A photograph of Hanson, Scott and betatron engineer Riesen with the scattering chamber is shown as Fig. 12. A photograph of the scattering chamber as modified for electron electron scattering is shown as Fig. 13. It shows the arrangement

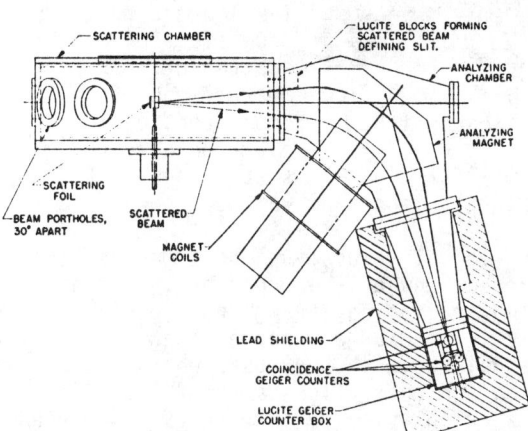

Fig. 11. Scattering chamber and detector arrangement.

Fig. 12. A. Hanson, M. Scott, and betatron engineer D. Riesen with the scattering chamber.

Fig. 13. Scattering chamber as modified for the electron electron scattering experiment.

of the target holder for selecting any of 10 scattering foils as well as a zinc sulphide screen and a vertical wire for checking the beam. The Faraday Cup in this case was inside the chamber. An energy spectrum of 15.7 MeV electrons scattered from polystyrene at 30° is shown in Fig. 14. The low energy group is from electron electron scattering. The measurement of the cross section for electron electron scattering was the subject of the thesis by Merrill Scott and was published as a separate paper.

In obtaining the experimental values for the cross sections for nuclear scattering, corrections of the order of 1 percent were made for multiple scattering, radiation straggling, double scattering and aperture size. The largest correction was that for radiative energy losses associated with the observed nuclear scattering. This correction for energy losses greater than 3 percent was obtained from a calculation by Schwinger. It varied from 5.4 to 9.1 percent. The average uncertainty due to counting statistics was about 1.2 percent. Upon including other experimental uncertainties the overall random errors were estimated to be 2 to 4%. The ratio of the experimental cross sections to the simple Mott formula is shown in Fig. 15 where it can be seen that the ratios increase systematically with Z. The theoretical values for the scattering by point nuclei were obtained with the help of interested theorists. The numerical evaluation of the cross sections for silver (Z=47) for our energy were made by Professor Feshbach. The values for gold were extrapolated to Z=79 and $\beta \approx 1$ from the results of Bartlett and Watson by Professor Bartlett who was at Illinois at that time.

The ratios of the observed cross sections to the calculated point charge cross sections are plotted in Fig. 16. The olid lines are the theoretical ratios calculated by Acheson for nuclear radii given by $R = 1.45 \, A^{1/3} \, 10^{-13}$ cm in terms of the change in a single phase shift in what corre-sponds to the S wave.[13] Upon comparing the experimental ratios

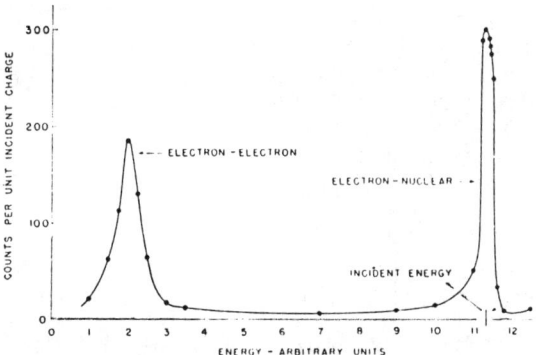

Fig. 14. Counting rate as a function of analyzer magnetic field for 15.7 MeV electrons scattered by polystyrene (carbon). The counters accept electrons which have lost up to 3 percent in energy. The ordinate does not represent the number of electrons per unit energy interval.

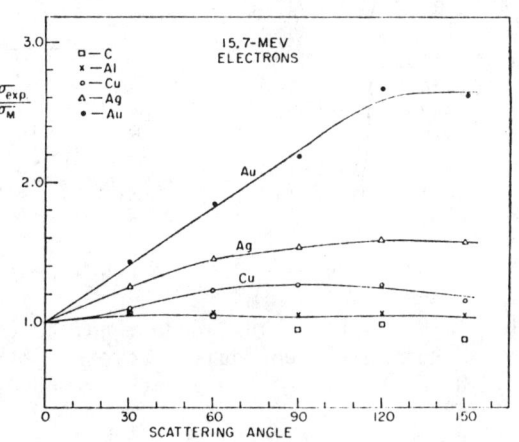

Fig. 15. Ratios of the experimental cross sections to the simple Mott formula. The lines sketched in to connect the points have no theoretical significance.

with the calculated values we reached the following conclusions. In the case of gold the results agreed fairly well with the assumption that the charge is distributed uniformly over the usually accepted nuclear volume. The agreement might have been somewhat better for a radius about 20 percent smaller.

This was a very comfortable conclusion and I am sure that it didn't change any ideas about nuclei and did not receive much notice. I do recall a one liner in one of the weekly magazines "Illinois group finds the heart of gold smaller than expected".

After reviewing the progress in electron scattering from the very early times to 1951, I became aware of the fact that for a brief period we had the highest energy electron beam in the world, high enough that the scattering was sensitive to the size of the nucleus. In that sense, we went a step "Beyond Mott Scattering". Other steps came quickly with Bob Hofstadter at leading the way at Stanford.

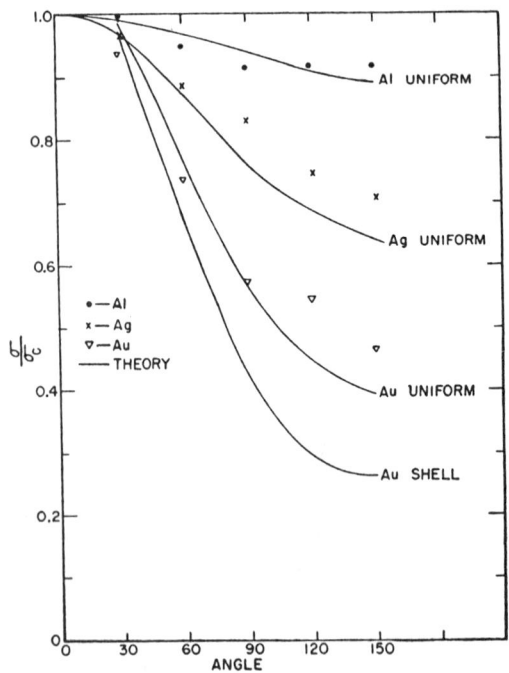

Fig. 16. Experimental cross sections shown as ratios to those for coulomb scattering. The lines represent the ratios calculated for nuclear radii given by $R = 1.45 A^{1/3}$ fm.

REFERENCES

1. E. Rutherford, The London, Edinburgh and Dublin Philosophical Magazine and Journal of Science, 6th series, 21, 669 (1911).
2. N. F. Mott, Proc. Roy. Soc. (London) A124, 426 (1929); A135, 429 (1932).
3. R. Hofstadter, Rev. Mod. Phys. 28, 214 (1956).
4. H. Feshbach, Phys. Rev. 88, 295 (1952).
5. H. V. Neher, Phys. Rev. 38, 1321 (1931).
6. A. Barber and F. C. Champion, Proc. Roy. Soc. A168, 159 (1938).
7. J. H. Bartlett and R. E. Watson, Proc. Am. Acad. Arts Sci. 74, 53 (1940).
8. N. F. Mott, Proc. Roy. Soc. A135, 429 (1932).
9. C. Shull, C. Chase and F. Myers, Phys. Rev. 63, 29 (1943).
10. H. Frauenfelder, et al.,. Phys. Rev. 106, 386 (1957).
11. R. J. Van de Graaff, W. W. Buechner and H. Feshbach, Phys. Rev. 69, 452 (1946).
12. E. M. Lyman, A. O. Hanson, and M. B. Scott, Phys. Rev. 84, 626 (1951).
13. L. K. Acheson, Jr., Phys. Rev. 82, 488 (1951).

A MICROSCOPE FOR NUCLEI AND NUCLEONS

Robert Hofstadter
Varian Laboratory of Physics
Stanford University, Stanford, California 94305

ABSTRACT

This paper discusses the early history of the electron scattering program and illustrates results concerning nucleons and nuclei.

In a historical article[1] on electron scattering in the decade of the fifties which I wrote in May 1985, I referred to the work of the Illinois group in the following quotation:

> "At the same time I tried to develop the electron scattering tools we would need to study nuclear structure. The large NaI(Tl) crystals could not be used because of pile-up due to the extremely low duty cycle of the linear accelerator and so I followed the early work of Lyman, Hanson and Scott at 15.7 MeV at Illinois by using a magnetic spectrometer. I needed a very good double focusing magnetic spectrometer and so I used the 180° design of the Cal Tech nuclear group. This kind of detector could perform well even with the terrible duty cycle of the linear accelerator."

As one can see from the quotation, I borrowed heavily on previous designs and studies. In particular, the Illinois group was extremely gracious. I have no recorded notes to refer to, but I remember how generous the group was when I made a trip to Champaign, probably in 1950 or 1951.

I had made the decision to study electron scattering from nuclei with the Stanford linear accelerator, and as I stated above, this required magnetic analysis. I remember how the Lyman, Hanson and Scott team shared information with me on their early results[2] at 15.7 MeV, and how they offered valuable advice on how I might deal with the higher electron energies soon to be made available at Stanford. I have always appreciated the very kind treatment given me at Illinois which enabled Stanford to start off in this new field with great confidence. I owe much also to Geoff Ravenhall, now a long-time Illinois faculty member, who was with us in the early days when he and Don Yennie made the relevant phase-shift calculations[3] providing accurate theoretical underpinnings for our experimental findings. Without their results the calculations based on the Born approximation would have given invalid interpretations of the data.

The so-called microscope which we used at Stanford in our initial work is sketched[4] in Figure 1. Figure 2 shows a photograph of the magnetic spectrometer mounted on an obsolete naval gun mount, which I obtained through the help of the Office

© American Institute of Physics 1987

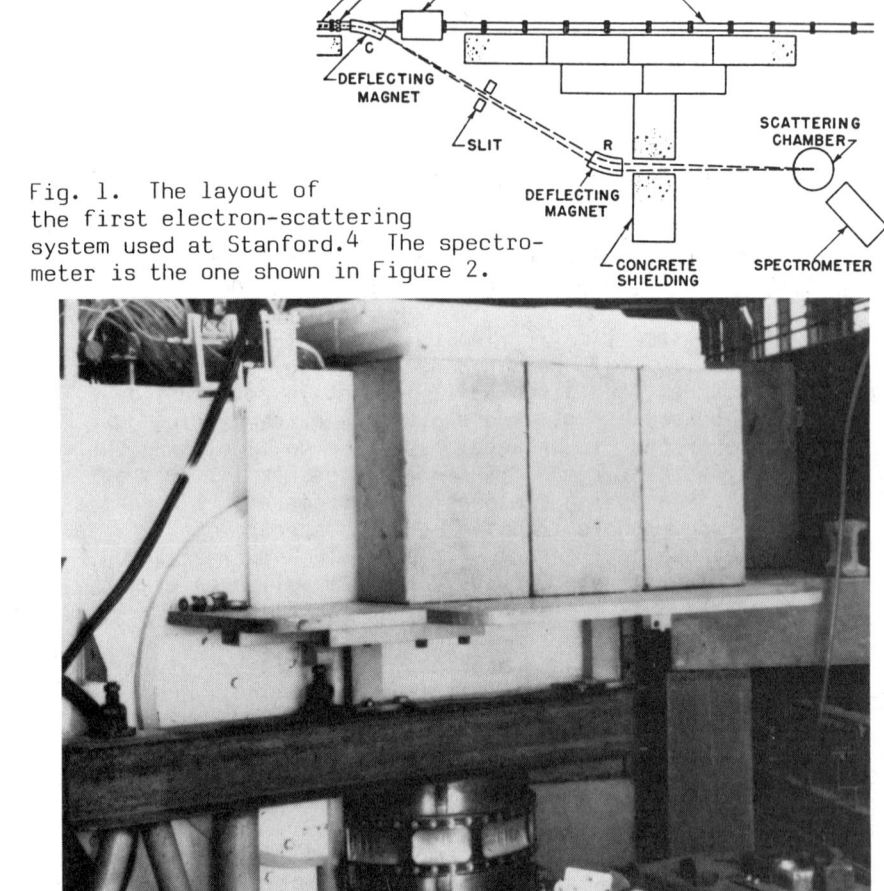

Fig. 1. The layout of the first electron-scattering system used at Stanford.4 The spectrometer is the one shown in Figure 2.

Fig. 2. The double focusing magnetic spectrometer is shown on its gun mount. The target chamber and monitor are also shown.

of Naval Research, which incidentally supported the electron scattering project for many years.

With this early system the Stanford group observed nuclear recoil peaks corresponding to separate isotopes in the target and, as shown in Figure 3, other inelastic peaks accompanying the elastic scattering peaks.[5,6] The latter paper[6] on beryllium has given rise to a whole industry of research on the energy levels of nuclei. While the elastic scattering gives a microscopic "picture" of the ground state of a nucleus, the inelastic scattering data shows details of the configurations of the excited states of nuclei with exquisite clarity. Indeed, if there were no other method of studying nuclear structure and nuclear excitations, the electron scattering method might yield a great deal of what is known all together about nuclei.

Some early elastic data on gold are summarized in Figure 4, which also shows solid lines representing the theoretical calcula-tions by Ravenhall and Yennie.[7] The experimental data of Figure 4 and their analysis resulted in the early so-called Fermi model of nuclear charge distributions of the type shown in the insert of Figure 6.

In lighter nuclei, such as oxygen-16, one could see pronounced "diffraction" features such as the dip indicated in Figure 5a. In subsequent investigations it was possible to see not only the first diffraction dip, as

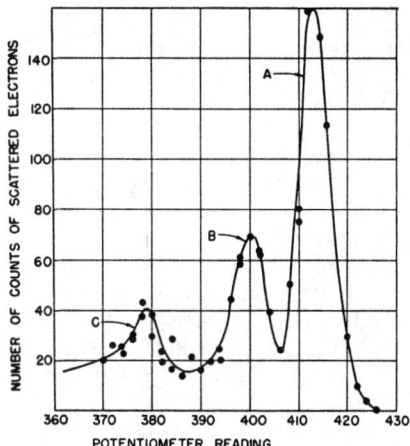

Fig. 3. A, B, and C indicate the elastic peak and the first two inelastic peaks in beryllium. This figure is taken from ref. 6.

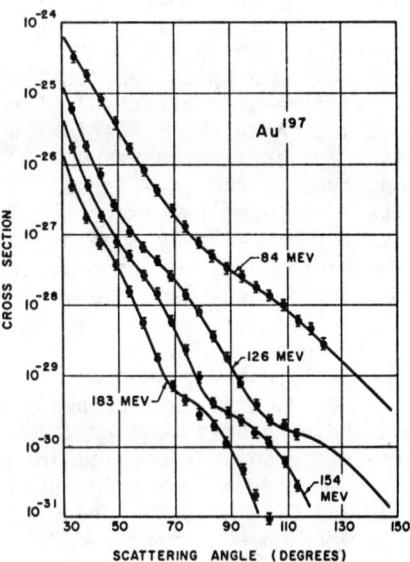

Fig. 4. Scattering curves in gold 197 and theoretical fits to the data.

shown in Figure 5a, but also the second dip as seen in the form factor curve of Figure 5b, where observations at different incident energies have enabled us to derive the ^{16}O form factor. Even higher-order dips were also observed in heavier nuclei.

Following a period in which many nuclei were studied, it became possible to summarize the ground state charge distributions for various spherical nuclei studied by the electron scattering method. The summary[8] is shown in Figure 6, and included the approximate charge distribution of the proton, which we shall now discuss.

The success of the electron scattering method in demonstrating finite size of nuclei, as first pointed out by Lyman, Hanson and Scott,[2] was followed up by our looking for finite size effects in the proton. An elastic scattering peak for the proton was observed in the very first Stanford paper on electron scattering, already mentioned as reference 5. In this paper the proton peak was observed and easily separated from the carbon peak in CH_2 (polyethylene) because of nuclear recoil. In following up these results, the proton and deuteron were studied in much more detail by using both CH_2 and CD_2 targets. Graduate student Bob McAllister and I made the proton our principal object of study.[9] In this paper we used a gaseous hydrogen target so that the finite size searched for could be observed as clearly as possible. The results

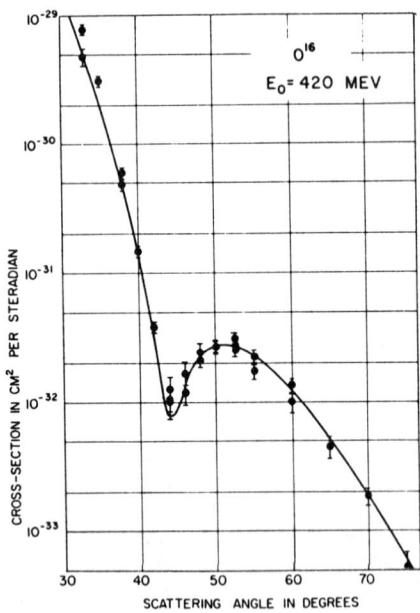

Fig. 5a. The differential cross section for ^{16}O observed at 420 MeV.

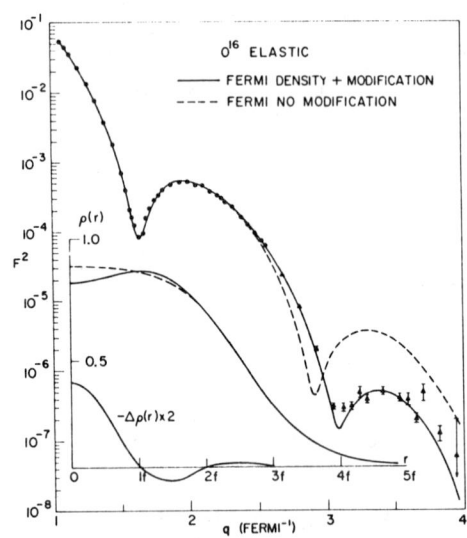

Fig. 5b. The ^{16}O form factor. Slightly different charge distributions produce different form factors.

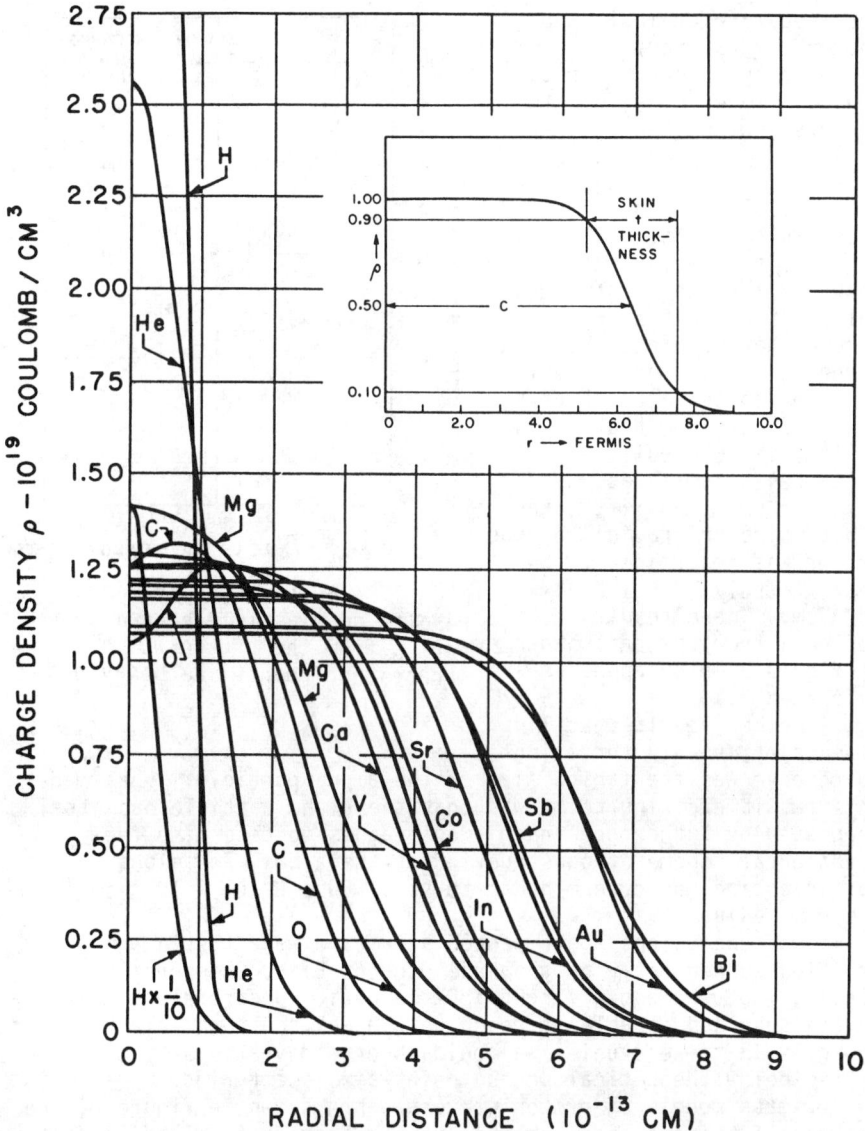

Fig. 6. The approximate shapes of the charge distribution of selected nuclei, including the proton and the alpha particle. Note the change of scale for the proton. The insert explains the Fermi model of spherical nuclear charge distributions in the ground state. The quantity "c" is the distance to the half-density point and "t" is the skin thickness, equivalent to the

obtained were dramatic and showed several very interesting things. First, finite size effects, or "form factors", were observed because the scattering cross section fell below the point size calculations of Rosenbluth.[10] Second, magnetic scattering was observed for the first time. Third, it was necessary to assume that both electric and magnetic proton models would require finite size. Fourth, it could be shown that the Coulomb law was valid at distances as small as 7×10^{-14} cm. And fifth, the finite size or "radius" of the proton was determined to be approximately $(7.4 \pm 2.4) \times 10^{-14}$ cm. These results followed from the data shown in Figure 7 which appeared in reference 9.

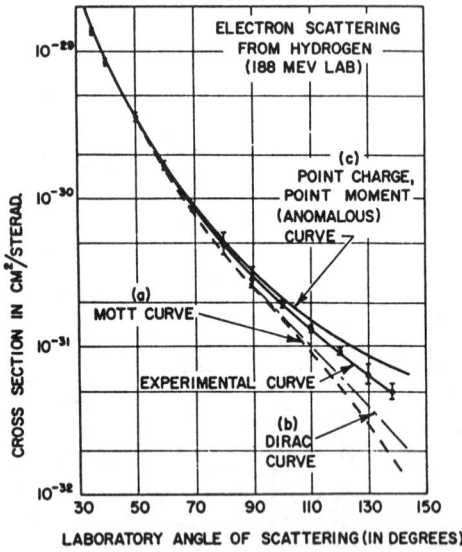

Fig. 7. The first experimental results on elastic electron scattering from the proton which demonstrated non-point-like behavior.

Another result that I found particularly thrilling to observe was the finite size of the alpha particle. What made this result exciting to me was that the alpha particle had itself been used by Rutherford and his colleagues to determine the fundamental scheme of construction of the atom. The alpha particle size was determined[11] to be $(1.60 \pm 0.10) \times 10^{-13}$ cm. for the rms radius.

The conclusions about proton structure were tested and verified subsequently by graduate student E. E. Chambers[12] and myself, who used a new larger 180° magnetic spectrometer, weighing thirty tons and capable of reaching 550 MeV scattered electron energy. In these studies we could observe deviations by a factor of ten below theoretical proton point-size scattering. Approximate models of proton structure are shown in Figure 8, and the best-fitting case in the figure corresponds to a "hollow" exponential structure. However, an exponential model of proton structure in which the charge density ρ is given by

$$\rho = \rho_0 e^{-r/a} \qquad (1)$$

fits nearly as well when the rms radius is chosen to be 0.80×10^{-13} cm. Note that in this figure the quantity $4\pi r^2 \rho$ is plotted as ordinate.

The Rosenbluth formulation for proton elastic scattering is given in equations 2-4.

$$\frac{d\sigma}{d\Omega} = \sigma_{NS} \{F_1^2 + \frac{\hbar^2 q^2}{4M^2 c^2} [2(F_1 + \kappa F_2)^2 \tan^2 \frac{\theta}{2} + \kappa^2 F_2^2]\} \qquad (2)$$

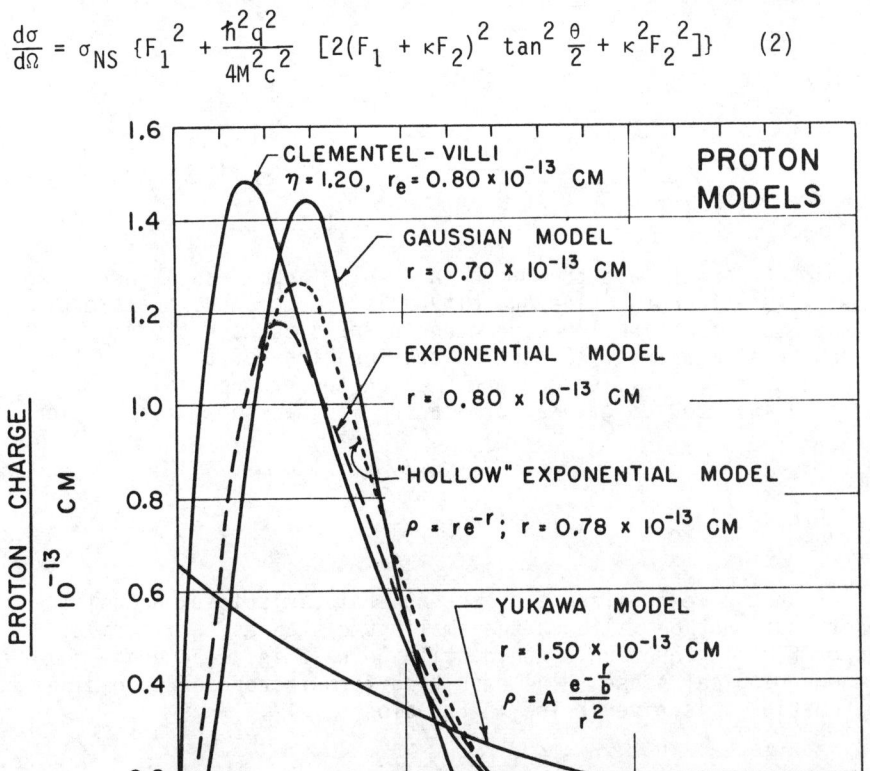

Fig. 8. Models of proton structure. The ordinate is $4\pi r^2 \rho(r)$.

with $\sigma_{NS} = \left(\frac{e^2}{2E}\right)^2 \frac{\cos^2\frac{\theta}{2}}{\sin^4\frac{\theta}{2}} \frac{1}{1 + \frac{2E}{Mc^2}\sin^2\frac{\theta}{2}}$ (3)

and $q = \frac{(2E/\hbar c)\sin\frac{\theta}{2}}{\sqrt{1 + (2E/Mc^2)\sin^2\frac{\theta}{2}}}$ (4)

in which F_1 and F_2 are the Dirac and Pauli (anomalous magnetic moment) form factors. The quantity q is the momentum transfer, M is the mass of the nucleon and other quantities have their conventional meanings. F_1 and F_2 are functions of the variable q.

In a non-relativistic approximation the exponential model could be inverted to give a form factor F(q) as shown in Equation 5.

$$F(q) = \frac{4\pi}{q} \int_0^\infty \rho(r) (\sin qr) r \, dr$$ (5)

Although we realized that the non-relativistic exponential model would not be valid at the smallest distances, the form factor F(q) in Equation 5 could still be used in an accurate phenomenological sense. The resulting form factor corresponding to Equation 1 is given below in Equation 6:

$$F(q) = \frac{1}{\left(1 + \frac{q^2 a^2}{12}\right)^2},$$ (6)

where a is the rms radius. This model was subsequently called a "dipole" model although it seems to me that the "exponential" model name, orginally used, should be kept. Nevertheless, the actual form factor, F, of Equation 6 has now survived for the last 30 years and still fits the data, even at much higher values of q.

To study the neutron it was clear that the deuteron problem should be attacked, since the deuteron was composed of one proton and one neutron. In some sense, if the deuteron's nuclear structure could be understood, then the effect of the proton's finite size could be allowed for and the neutron's "size" or form factors could be determined. These ideas were well understood by L. I. Schiff and by Yennie, Levy, and Ravenhall.[13] V. Z. Jankus, who was Schiff's graduate student at the time, made electron deuteron calculations[14] in anticipation of future elastic and inelastic electron scattering studies of the deuteron. Experimentally, M. R. Yearian and I made the first studies[15] of the magnetic form factors of the neutron. This work was also related to the earlier experimental findings[16] of J. A. McIntyre who studied the elastic scattering of electrons from the deuteron. With McIntyre's elastic deuteron factors (nuclear form factors) and Jankus' inelastic calculations it became possible to determine approximate form factors of the neutron. Now the "inelastic continuum", calculated by Jankus for the deuteron, had already been observed by us in the alpha particle[8] and represented the combined scattering from protons and neutrons in motion in the Fermi sea. In the deuteron the situation was even simpler than in the alpha particle since only one proton and one neutron were in relative motion, and this "motion" could be described in principle through the findings of McIntyre and previous studies of the deuteron by more conventional methods.

In Reference 8, I had suggested a rough method, the "area method", of finding the neutron's form factors by assuming incoherence in the electron inelastic scattering (inelastic continuum) from the deuteron, which implied the electro-disintegration or breakup of the deuteron. An example of the inelastic continuum of the deuteron is shown in Figure 9. From data of this kind it is possible to use both the area method and the "peak method" to determine the form factors of the neutron.

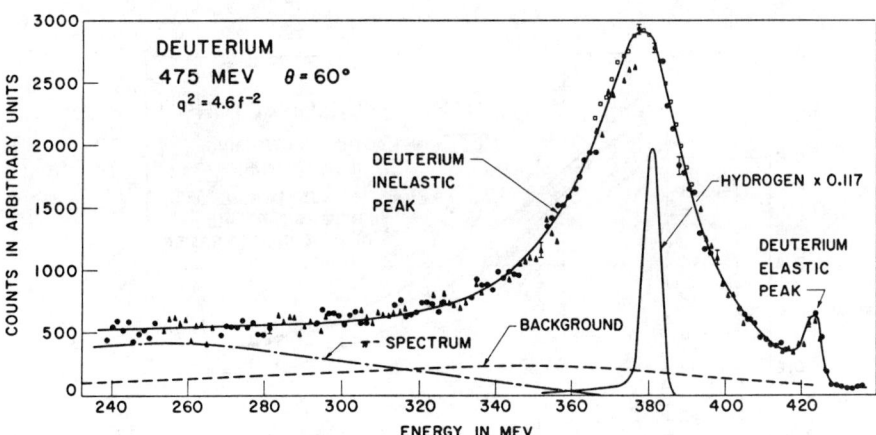

Fig. 9. In this figure are shown both the inelastic continuum of the deuteron, as well as the elastic deuteron peak, and also the corresponding free proton peak. A pion background is also shown, arising from electropion production.

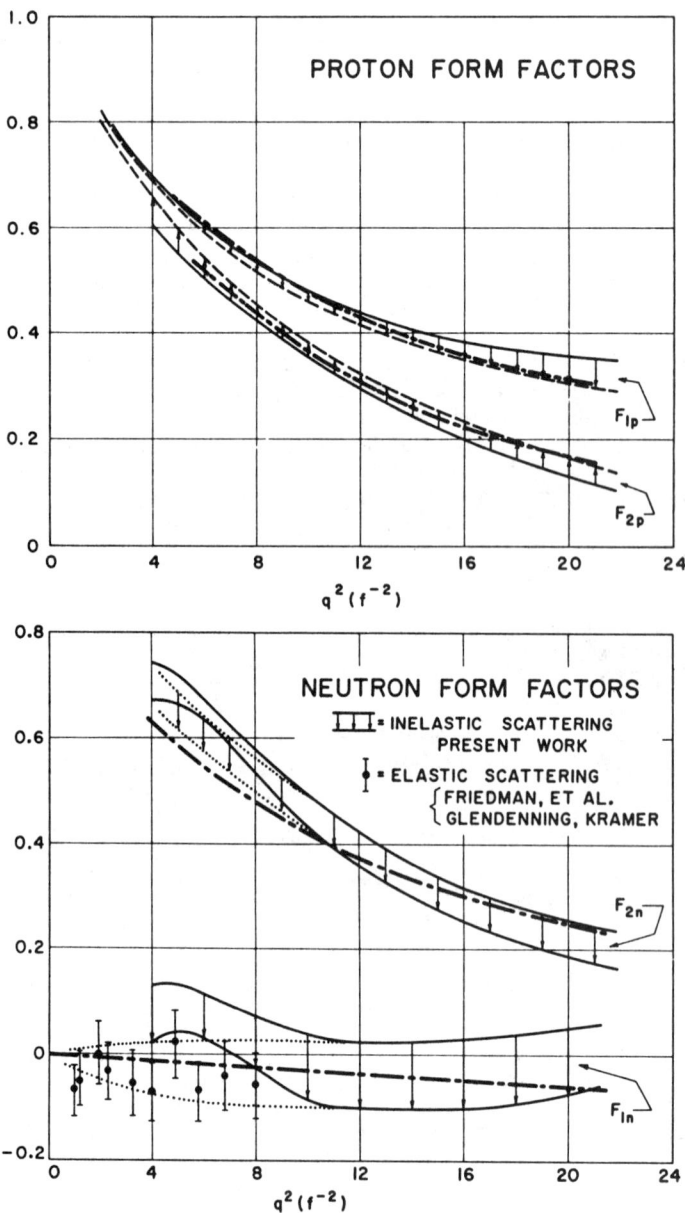

Fig. 10. F_1 and F_2 for both proton and neutron.

Jankus' calculations were improved with relativistic corrections by L. Durand[17] so that the "peak method" as well as the area method could give reliable determinations of neutron form factors. In this work on the deuteron the results showed that the magnetic form factor of the neutron satisfied the dipole model with an rms radius lying between 0.8 and 0.9 x 10^{-13} cm. And furthermore, the behavior of the two constituents of the deuteron, proton and neutron, could be clearly observed in the momentum distribution of each nucleon in the deuteron's bound state.

It is worth emphasis to point out that the electron scattering method uncovered the physics of the nucleons in the deuteron in a very clear fashion so that the interpretation of the form factors was quite clean. Similar work could be done in the inelastic continua of the alpha particle and the beryllium nucleus and both yielded results agreeing with those observed in the deuteron. With results coming from both nuclei and nucleons we see an excellent example of how the phenomena of nuclear physics and elementary particle physics were being revealed at the same time by the electron scattering method. In other words, the electron scattering process provided an ultramicroscope for both nucleons and nuclei.

One important prediction was made by Y. Nambu[18] who showed that the proton and neutron form factors could be explained simultaneously by assuming the existence of a new heavy neutral meson. This meson is now called the omega meson and was discovered soon afterwards at Berkeley[19] where it was observed that this meson decayed into three pions when produced in proton-antiproton collisons at 1.61 GeV/c. Many other theoretical developments followed from our studies of the nucleon form factors. I have outlined some of these developments in the historical article[1] referred to at the beginning of this paper.

The proton and neutron form factors and "sizes" were set out by the early work of the Stanford group in the 1950's. More detail was obtained in the early sixties by using a new, larger, magnetic spectrometer weighing 200 tons with analyzing ability reaching 1000 MeV/c. A photograph of the 1000 MeV/c spectrometer, along with the earlier 550 MeV/c spectrometer was given in Reference 1. Some of our last data on neutron and proton scattering are shown in Figures 10 and 11. These were obtained with the 1000 MeV/c spectrometer.

This article would not be complete without mention of the very recent determination[20] of form factors covering q^2 regions 12 to 25 times those corresponding to our data of the fifties and sixties. The data of Reference 20 are shown in Figure 12. The scale is to be noted in this figure because it extends up to 12 $(GeV/c)^2$ whereas our old data stopped at 1 $(GeV/c)^2$. The form factors in this figure are expressed in terms of a product of $(q^2)^{n-1}$ and the square root of the ratio of the cross sections to the Mott (σ_{NS}) cross section. This type of plot reveals deviations from q^4 behavior which are equivalent to a choice of $n = 3$. The power law $n = 3$ corresponds exactly to the extension of our old dipole form factor, and theory indicates that this

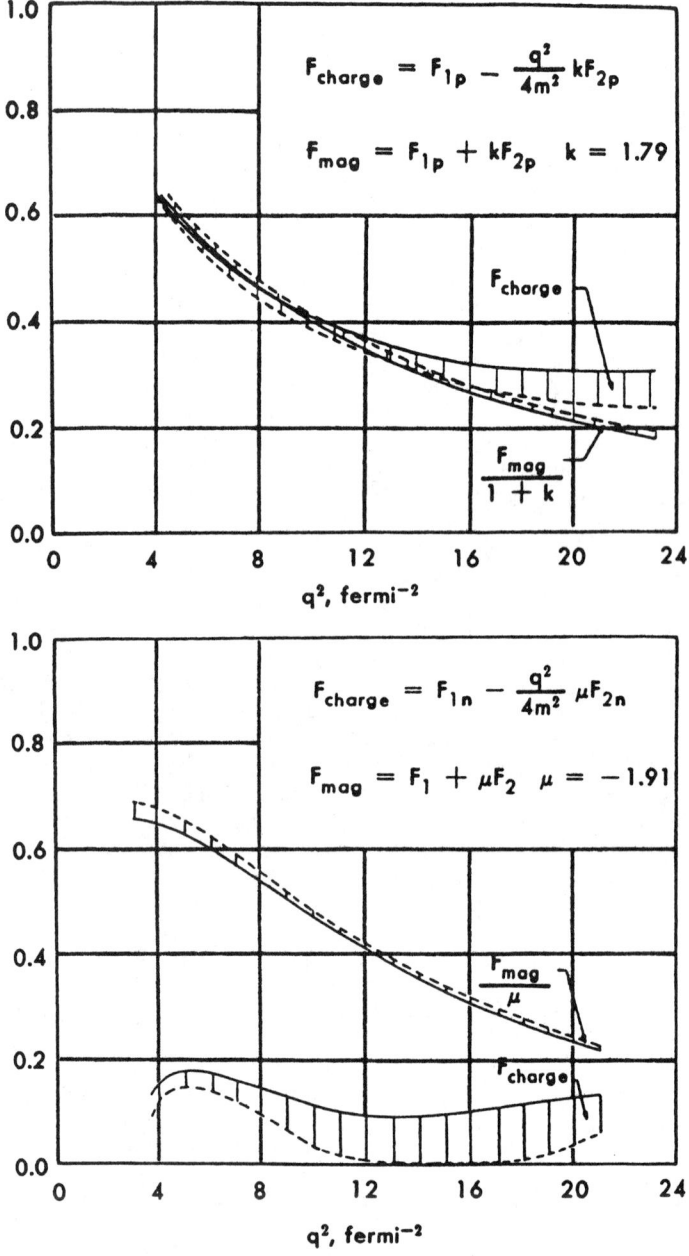

Fig. 11. The "modified" form factors, F_{charge} and $F_{magnetic}$, for the proton and the neutron.

quantity gives the number of elementary quark constituents of the nucleon. The deviations from the dipole model are small and provide information on the details of the quark-gluon vertex and the point-like characters of quark and their interactions with each other. I find it remarkable that our old exponential model, that is, the dipole model has held up so well 30 years later.

Our Stanford electron scattering data were taken at a time that was fortunate for me and my colleagues. We had the complete decade of the fifties, in which I, my graduate students, and my colleagues could find so many of the fundamental properties of nucleons and nuclei. Our groups were small, rarely exceeding three individual physicists. These small groups designed and built the apparatus, and comfortably took the data that were sought. In simple terms, we had a monopoly in the field of nucleon and nuclear structure mostly because we had an excellent linear accelerator conceived by W. Hansen before his untimely death.

Bill Hansen may or may not have known how fine a microscope he had designed. But together with L.I. Schiff, F. Bloch, and the Illinois group they had all provided the means by which my colleagues and I could obtain once-in-a-lifetime experimental results. My profound thanks go to them for giving me this opportunity.

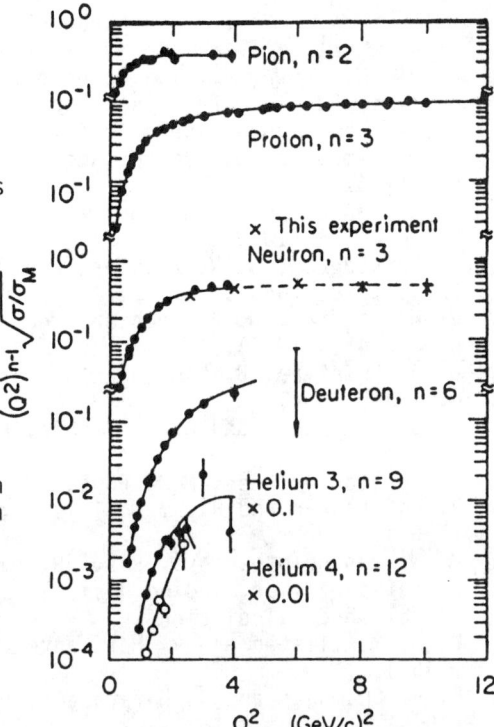

Fig. 12. The recently determined[20] magnetic form factors of the proton and neutron.

REFERENCES

1. Laurie M. Brown, Lillian Hoddeson, and Max Dresden, Pions to Quarks: Particle Physics in the 1950s (Cambridge University Press), to be published.
2. E. M. Lyman, A. O. Hanson, and M. B. Scott, Phys. Rev. 84, p. 626 (1951).
3. D. R. Yennie, R. N. Wilson, and D. G. Ravenhall, Phys. Rev. 92, p. 1325 (1953).
4. R. Hofstadter, H. R. Fechter, and J. A. McIntyre, Phys. Rev. 92, p. 978 (1953).
5. R. Hofstadter, H. R. Fechter, and J. A. McIntyre, Phys. Rev. 91, p. 422 (1953).
6. J. A. McIntyre, B. Hahn, and R. Hofstadter, Phys. Rev. 94, p. 1084 (1954).
7. D. G. Ravenhall and D. R. Yennie, Phys. Rev. 96, p. 239 (1954).
8. R. Hofstadter, Rev. Mod. Phys. 28, p. 214 (1956).
9. R. Hofstadter and R. W. McAllister, Phys. Rev. 98, p. 217 (1955).
10. M. N. Rosenbluth, Phys. Rev. 79, p. 615 (1950). L. I. Schiff had also earlier studied deviations from point size in an unpublished calculation.
11. R. W. McAllister and R. Hofstadter, Phys. Rev. 102, p. 851 (1956).
12. E. E. Chambers and R. Hofstadter, Phys. Rev. 103, p. 1454 (1956).
13. D. R. Yennie, M. M. Levy, and D. G. Ravenhall, Rev. Mod. Phys. 29, p. 144 (1957).
14. V. Z. Jankus, Phys. Rev. 102, p. 1586 (1956).
15. M. R. Yearian and R. Hofstadter, Phys. Rev. 110, p. 552 (1958).
16. J. A. McIntyre, Phys. Rev. 103, p. 1464 (1956).
17. L. Durand III, Phys. Rev. Letters 6, p. 628 (1961).
18. Y. Nambu, Phys. Rev. 106, p. 1366 (1957).
19. B. C. Maglic, L. W. Alvarez, A. H. Rosenfeld, and M. L. Stevenson, Phys. Rev. Letters 7, p. 178 (1961).
20. S. Rock, R. G. Arnold, P. Bosted, B. T. Chertok, B. A. Mecking, I. Schmidt, Z. M. Szalata, R. C. York, and R. Zdarko, Phys. Rev. Letters 49, p. 1139 (1982).

SESSION B

ELECTRON SCATTERING AND FEW-NUCLEON SYSTEMS

Bernard Frois
Service de Physique Nucléaire - Haute Energie
CEN Saclay, 91191 Gif-sur-Yvette Cedex, France

It is really wonderful to be here with so many friends to celebrate the first electron scattering experiment in nuclear physics which was performed here at the University of Illinois thirty-five years ago. This pioneering work was the dawn of a new era. Electron accelerators have since played a central role in the development of nuclear and particle physics.

French physicists have rapidly realized the advantages of electromagnetic interactions to probe the structure of matter. They have developed strong collaborations with their colleagues in the United States. Among these cooperative efforts the Illinois-Saclay collaboration has been a very successful one. Peter Axel, who unfortunately has now left us, has largely contributed to this success by encouraging extensive exchanges between our laboratories. I am in particular very indebted to him for his constant interest and his warm encouragement. So before beginning my talk I would like today with all my colleagues from Saclay to wish a very happy anniversary to the University of Illinois and I hope that this very special link will last for many years.

1. INTRODUCTION

The solution of the nuclear many-body problem is still at present beyond the reach of theory. It is only for few-nucleon systems, A=2 and A=3, that one has found exact solutions of the Schrödinger equation in terms of a realistic nucleon-nucleon interaction. Thus the study of few-nucleon systems is essential in order to explore the limits of the mesonic description of nuclear interactions. The electromagnetic properties of the two- and the three-nucleon systems are of special interest. In particular the elastic form factors of the deuteron, and of the tri-nucleon systems are a rich testing ground for theory. After the pioneering work done at Stanford, considerable experimental difficulties had to be overcome to extend the measurements up to sufficiently high momentum transfers. It is only recently that all these form factors have been measured with precision up to 1 $(GeV/c)^2$.

The proceedings of the conference of Los Angeles in 1972 on the few-body problem show that the first generation of experiments had revealed the basic challenges. The main difficulty was that data were difficult to interpret because there were no complete calculations. It is amusing to realize today that the Los Angeles conference was going to be the start of the present generation of experiments and calculations. One was aware that the central problem was to determine to what limit it is possible to describe the structure of nuclei in an approach based only on nucleons and two-body interactions.

2. MESON-EXCHANGE CURRENTS

The importance of meson-exchange currents in electromagnetic data was realized shortly after the pion was discovered. Such currents are required by electromagnetic gauge invariance as well as chiral symmetry. However, the large coupling constants associated with meson-nucleon vertices do not allow the derivation of a convergent diagrammatic expansion. The inclusion or omission of specific diagrams changes the interpretation of experimental data.

Chemtob and Rho[1] by highlighting the role of chiral symmetry have identified a clear hierarchy of dominant processes where the pion plays a central role. The description of the π exchange current is constrained by model-independent theorems, valid for "soft pions" which have a small momentum at the scale of the nucleon mass. The first investigations of the effects of mesons in nuclei have relied on these model independent predictions. The classical example is the slight increase observed in the thermal neutron capture by the deuteron. Riska and Brown[2] have shown that the 10 % disagreement between experiment and theory was the signature of the π exchange current. Various experimental results have now clearly demonstrated that low energy theorems are operative in nuclei[3].

It is crucial to investigate the validity of mesonic theory as a function of momentum transfer, to find the momentum at which it begins to break down. This is precisely what has been accomplished by electron scattering experiments in the last decade.

3. ELECTRODISINTEGRATION OF THE DEUTERON AT THRESHOLD

This M1 isovector transition is an admixture of two amplitudes, the $^3S_1$ and the $^3D_1$ components of the ground state wave function of the deuteron coupled to the $^1S_0$ state of the n-p system. It is the inverse reaction of the neutron capture $n + p \rightarrow D + \gamma$. In thermal neutron capture, the pion has negligible momentum and the contribution of the π-exchange current is given in a model-independent way by low-energy theorems. In the electrodisintegration of the deuteron at threshold, the spatial distribution of meson-exchange currents can be explored with virtual photons of adjustable wavelength. The nucleonic and mesonic currents which contribute to the cross section have strong destructive interferences which occur successively at different momentum transfers. Thus, measurements at specific momentum transfers isolate the contributions from different meson exchange processes.

Cross sections for this reaction have been measured up to 28 fm^{-2}. Experimental data[4], averaged over the energy of the n-p system near threshold ($E_{np} \leq 3$ MeV) are shown in figure 1 together with theoretical predictions which use the Paris potential[5]. The predictions take into account both the effect of both nucleons and mesons. The purely nucleonic contribution has a deep minimum around

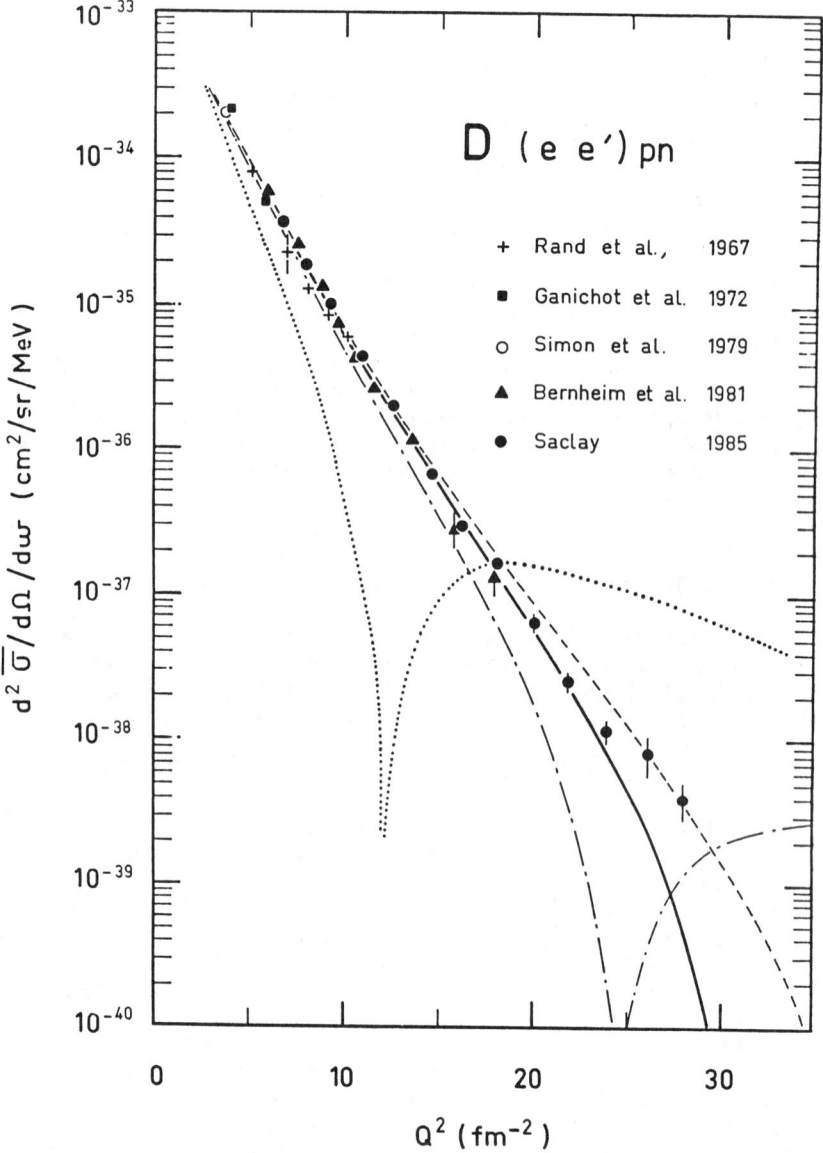

Fig. 1 - Cross sections for the electrodisintegration of the deuteron at threshold. The curves represent the contributions of nucleons (dotted), nucleons + pions (dash-dotted), nucleons + pions + ρ-mesons (dashed), nucleons + pions + ρ-mesons + Δ (solid). Hadronic form factors have been included.

$Q^2 = 12$ fm^{-2} resulting from a destructive interference between the $^3S_1$-$^1S_0$ and the $^3D_1$-$^1S_0$ amplitudes. Non-nucleonic degrees of freedom are essential for the interpretation of the data[5,6]. In a large momentum transfer range, between 10 and 15 fm^{-2}, they account for nearly 100 % of the experimental cross section. This process provides some of the most striking evidence of the presence of meson-exchange currents in nuclei.

The experimental data are well described by theoretical predictions which allow for π, ρ and Δ meson-exchange currents. Up to $\sim$ 15 fm^{-2} there is almost no sensitivity to the choice of the nucleon-nucleon potential or the πNN form factor. The theoretical predictions for higher momentum transfers strongly depend on the detailed structure of the currents and wave functions.

In order to achieve the best description of the data beyond 10 fm^{-2}, one must include the effects of the π, ρ and the Δ meson exchange currents with hadronic form factors at the meson nucleon vertex. The Dirac form factor F_1 must be used to fit the deuteron electrodisintegration data. The choice of the Sachs form factor G_E would lead to a large discrepancy with experiment. The finite size of the meson-nucleon coupling plays an important role at the spatial scale probed by high momentum transfers. This size is accounted for by a hadronic form factor at the meson-nucleon vertex, with a cutoff parameter $\sim$ 1200 MeV in the πNN form factor. This corresponds to a size of 0.5 fm for the hadronic interaction region. This value, which is smaller than the proton charge radius, appears to be consistent with the description of the nucleon in a two-phase model where charge and baryon number are fractioned between a quark core and a pion cloud[7]. A radius of 0.54 fm is predicted for the core while the radius of the cloud which includes the ρ-meson propagation is 0.91 fm. The difference between the electromagnetic and the hadronic form factors comes from the difference in the coupling of the photon and the pion to the nucleon. The photon, because of its large coupling to the ρ-meson, is sensitive to the cloud, while the pion is partially blind to the cloud and probes essentially the size of the nucleon core.

In order to avoid the use of phenomenological hadronic form factors, a different approach has been recently proposed[8] in which meson-exchange currents are derived from the nucleon-nucleon potential through the continuity equation. The advantage of this procedure is that hadronic form factors are automatically consistent with the nucleon-nucleon potential. This procedure yields very similar results to the perturbative approach. Therefore one is quite confident of the reliability of the mesonic description of the electrodisintegration of the deuteron at threshold.

The ρ-meson exchange and the Δ-isobar current tend to cancel the effect of the πNN form factor. Short range processes tend to cancel each other in this isovector process, and it seems that the only really significant contribution is due to the π-exchange current between two-nucleons in a pointlike coupling[9]. It is quite intriguing that the only term which survives is the one constrained by chiral symmetry even at such large momentum transfer (figure 2). This result suggests that chiral symmetry plays an important role for the description of meson-exchange currents beyond the description of soft pions.

A few calculations have begun to investigate the role of quarks with phenomenological models[10]. In the momentum transfer range measured at present, the role of mesons still predominates, quarks having essentially a negligible contribution. Experimental data must

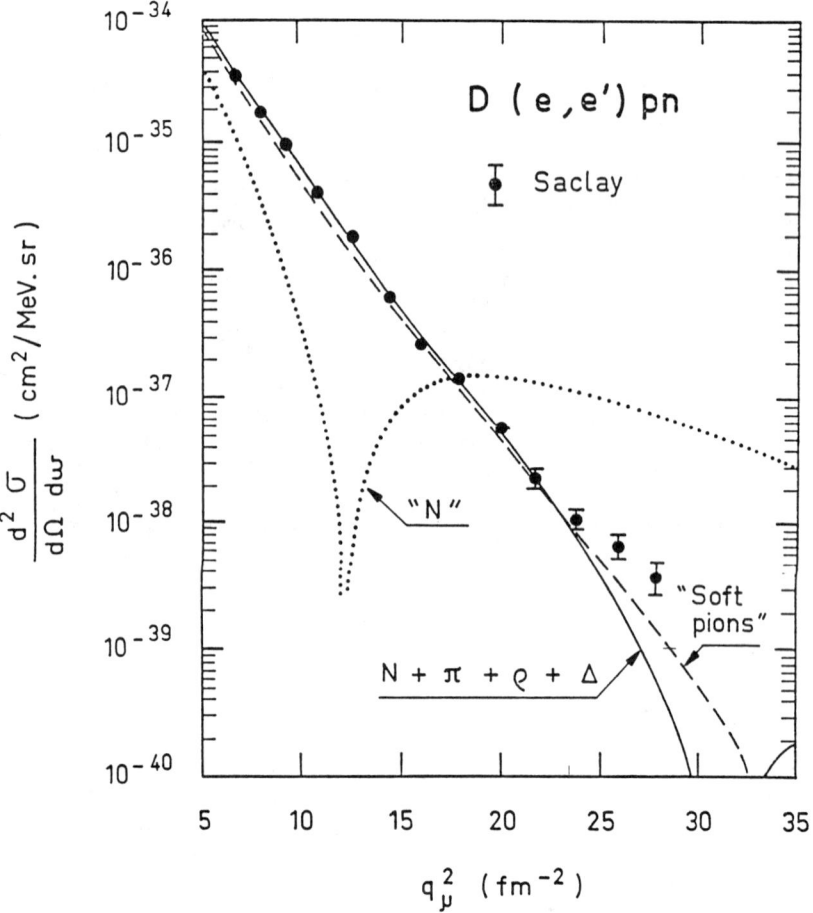

Fig. 2 - Electrodisintegration of the deuteron. The dotted curve is the predition with nucleons only. The dashed curve is the contribution of nucleons and pions with a point coupling. The solid curve is the full calculation with a meson nucleon interaction region of 0.5 fm.

now be extended to high momentum transfers to investigate such short range processes.

The most surprising result of this experiment is that there is not yet any sign of breakdown of mesonic theory, even at 1 $(GeV/c)^2$, beyond its expected limit of validity.

4. THE MAGNETIC FORM FACTOR OF THE DEUTERON

The magnetic form factor $B(Q^2)$ of the deuteron is of isoscalar nature. Therefore, the contribution of the pion-exchange current derived from chiral symmetry, which plays a central role in isovector processes, vanishes. Thus, one has a unique opportunity to study the effects of other meson-exchange currents. Recent experiments[11] have considerably extended the momentum transfer range of the measurements of $B(Q^2)$, providing a very discriminating test of theoretical calculations (figure 3).

Gari and Hyuga[12] have shown that the $\rho\pi\gamma$-exchange current which is the dominant isoscalar meson-exchange current, increases the impulse approximation cross section by a factor 3 at 30 fm^{-2}, in agreement with experiment. However, beyond 30 fm^{-2}, a smooth fall-off is predicted, while the experimental data show a minimum around 50 fm^{-2}.

The effects of $\Delta\Delta$ and NN*(1440) components in the ground state of the deuteron have been investigated with a coupled channel NN interaction by Sitarski, Blunden and Lomon[13]. They have used NN scattering data up to 1 GeV, the magnetic moment and the magnetic form factor of the deuteron as constraints. The effects of meson exchange, as calculated by Gari and Hyuga[12], have also been included, since they are assumed to be relatively independent on the details of the NN potential. This model predicts the existence of a diffraction minimum around 45 fm^{-2}. The best agreement with experiment up to 60 fm^{-2} is obtained with NN(D state) = 5.45 %, $\Delta\Delta \sim 1$ % and NN* ~ 0 %.

Meson exchange currents are of relativistic order for an isoscalar process. The data have reached very high momentum transfers and a non-relativistic framework may no longer be reliable for such a light system as the deuteron. At present, it is not yet possible to draw a reliable conclusion on relativistic effects. Various relativistic approaches[14] give completely different predictions. It is now crucial to investigate the origin of their discrepancies and to find the appropriate relativistic framework for the description of the properties of the deuteron.

The parton model does not predict a diffraction minimum[15]. The existence of a minimum around 50 fm^{-2} shows that even at such high momentum transfers asymptotic behavior has not been reached. Hybrid models[9,16] involving both nuclear and quark degrees of freedom are able to give a qualitative description of these data.

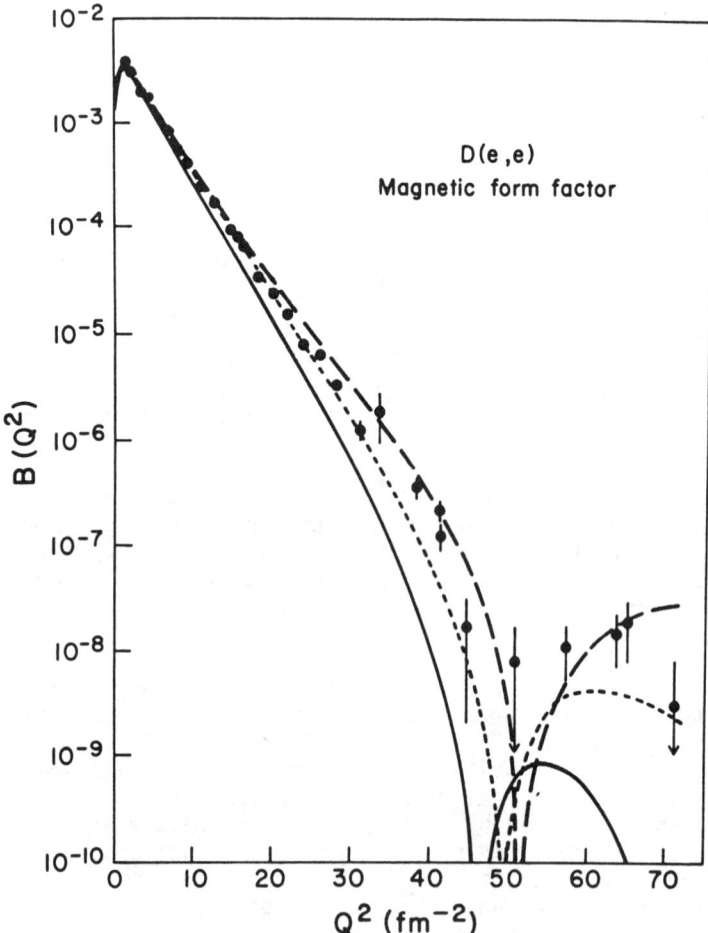

Fig. 3 - The magnetic form factor of the deuteron[11]. The thick curve is the impulse approximation. The short-dashed curve includes the effect of Δ-Δ components and meson-exchange currents[13]. The long-dashed includes the effect of meson-exchange currents calculated in the Skyrme model from the chiral anomaly[17].

Nyman and Riska[17] have studied the isoscalar electromagnetic current operator in the Skyrme model. They have shown that this current is determined by the part of the effective QCD Lagrangian that accounts for chiral anomalies and which is essentially model independent. The isoscalar meson exchange currents are uniquely related to the baryonic current. It is then possible to derive the form factors of the deuteron from the isoscalar electric structure of the nucleons themselves. This approach gives a good description of the experimental data (figure 3).

5. THE TRI-NUCLEON SYSTEMS

The simplicity of the deuteron makes it a natural starting point for the study of nucleon interactions in nuclei. The success of the Impulse Approximation corrected for meson exchange currents, in describing its properties naturally leads to asking whether this scheme can be as successful for more complex nuclei. In such systems short-range interactions play a different role ; the deuteron is barely bound (1 MeV/nucleon instead of the 8 MeV/nucleon typical of heavy nuclei) and the two nucleons rarely find themselves at close proximity. Moreover, the influence of many-body forces (3-body forces in particular) or isospin dependent effects cannot be examined. These aspects of the nuclear force can be studied in the tri-nucleon systems (^{3}He and ^{3}H).

Calculation of properties of A=3 nuclei with realistic forces is nowadays reliable. Hamiltonians based on realistic two-body interactions yield almost identical results for the binding energies of ^{3}H and ^{3}H which are significantly lower than the experimental values (7.72 and 8.49 MeV). The difference between experiment and theory for the binding energy of the A=3 system is now believed to arise largely from the three-nucleon force[18]. Only the long range (attractive) part of the three-body force which is attributed to two-pion exchange, is well understood. The short range part is believed to be repulsive and is adjusted phenomenologically. In a more ambitious approach[19] both the short and long range parts of the three-body force are derived microscopically by including meson exchange and isobar-degrees of freedom. These calculations render support to the phenomenological approach by producing similar trends. The effect of the three-body force is to increase the binding energy by 1-3 % of its potential energy. The resulting smaller rms radii are also in agreement with the experimental values. The detailed study of the charge and magnetic form factors of ^{3}H and ^{3}He can provide further insights on the structure and the momentum content of the ground state.

5.1 The Form Factors of ^{3}H and ^{3}He

Elastic electron scattering cross section from ^{3}H or ^{3}He is a linear combination of the squares of the charge and magnetic form factors. Recent measurements[20] have separated both form factors up to ~ 25 fm^{-2}. Figure 4 provides a comparison of the experimental data for the charge form factors to the theoretical predictions of Hajduk and Sauer[19]. The impulse approximation, which accounts only for nucleon currents does not describe the data correctly. The diffraction minimum is shifted by 3 fm^{-2}, while the amplitude of the second diffraction maximum is too low by a factor 2. Various combinations of two-body and three-body forces have been investigated, none of them is able to account for the experimental charge form factors. The influence of the three-body force on their shape is very small. A larger negative contribution which vanishes at q=0 and which increases up to the second diffraction maximum is needed. There is a term in the two-pion exchange three-nucleon interactions

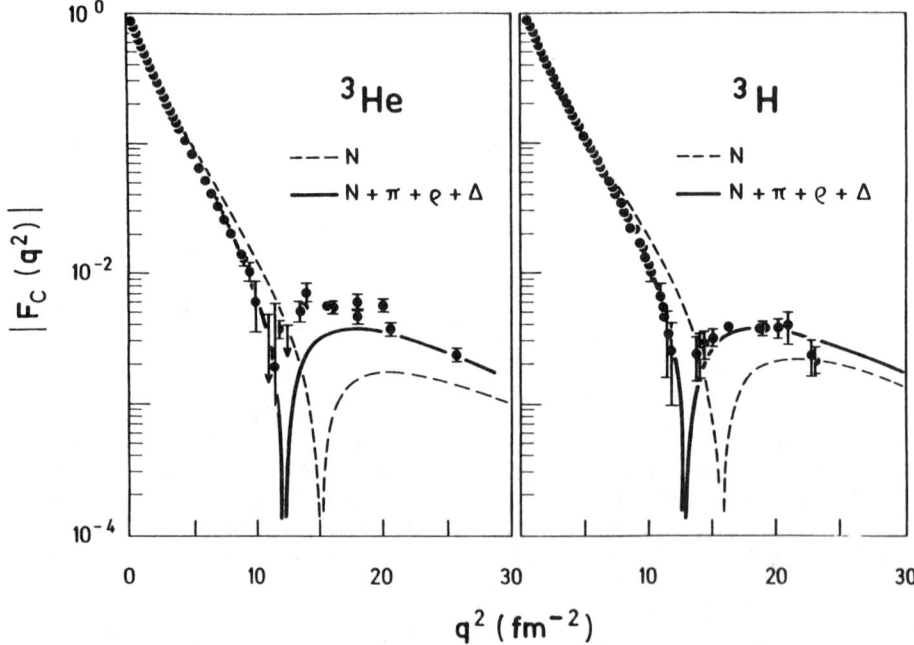

Fig. 4 - The charge form factor of ^{3}H and ^{3}He. The dashed curve corresponds to the impulse approximation. The solid curve is the result of the full calculation.

that gives such a contribution, but in order to explain the observed form factors its strength has to be increased beyond reasonable limits[18,19,21].

As in the case of the deuteron, the charge form factors of the three-nucleon system cannot be described without the explicit introduction of non-nucleonic degrees of freedom. A reasonable agreement is obtained for both form factors when the effects of Δ-isobars and meson-exchange currents are included[19] (figure 4).

The relative importance in a non-relativistic framework of the meson exchange currents depends on the choice of the pion-nucleon coupling. In the pseudoscalar coupling the isoscalar and the isovector meson exchange contributions have approximately the same magnitude. In ^{3}He the two contributions have the same sign, so the ^{3}He form factor is not very sensitive to the choice of coupling. In ^{3}H they contribute with opposite signs which leads to almost a complete cancellation for pseudoscalar coupling of the effect of meson exchange currents. Such a cancellation is not observed experimentally.

A description of the same data has also been done with quark models[22]. These models give results which are similar to the results obtained with the nucleon pair current with the pseudo-vector coupling. The photon is coupled to a quark-antiquark pair instead of the usual nucleon-antinucleon pair used for the pion exchange current. These calculations are in reasonable agreement with experiment, but have a large model dependence.

The major sources of uncertainty in the theoretical description of the tri-nucleon charge form factors are of relativistic origin. Meson exchange currents are relativistic corrections while the three-body wave functions are calculated in a non-relativistic framework. Unfortunately, at this point we lack the appropriate theoretical framework for incorporating in a consistent and complete fashion relativistic corrections.

The magnetic form factors of ^{3}H and ^{3}He [ref.20] are shown in figure 5. They correspond to an M1 isovector transition similar to the electrodisintegration of the deuteron. The impulse approximation alone cannot explain the experimental data. In the region of $q^2 = 8$ fm^{-2} for ^{3}He and 12 fm^{-2} for ^{3}H the cross section is entirely due to non-nucleonic processes because of destructive interference among the nucleonic amplitudes. The prediction of Strueve, Hajduk and Sauer[19] which accounts for both the effect of π and ρ meson exchange currents agrees with the experimental data up to the diffraction minimum. In the region of the second diffraction maximum, there is only a slight deviation from experiment. Similar calculations[21-23]

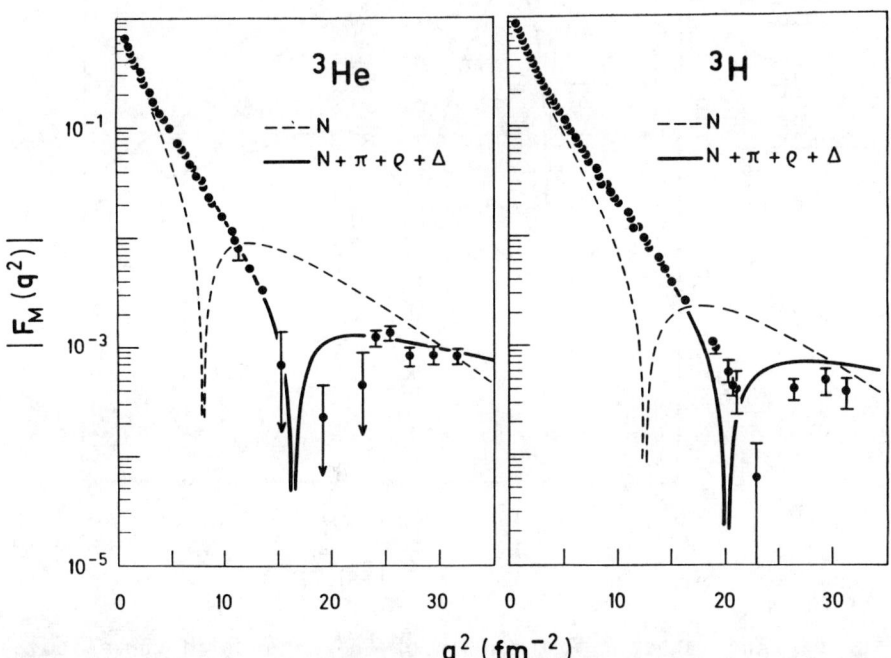

Fig. 5 - The magnetic form factor of ^{3}H and ^{3}He. The meaning of the curves is the same as figure 4.

are also in good agreement with the ^{3}He and ^{3}He magnetic form factors. Quite surprisingly the form factor of ^{3}He is well described by the impulse approximation and the "soft" pion contribution, exactly as for the electrodisintegration of the deuteron (figure 6). In the expression of the exchange current operators, one must use the Dirac form factor F1 to reproduce the experimental data, while there is strong discrepancy at large momentum transfers when the Sachs form factor G_E is used.

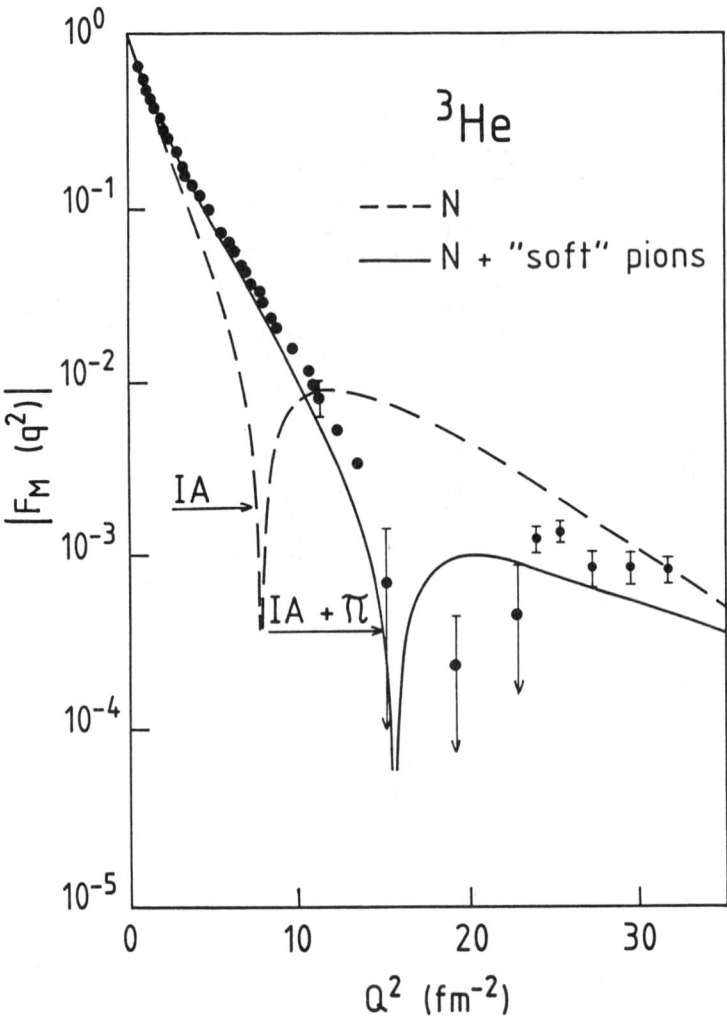

Fig. 6 - The magnetic form factor of ^{3}He. The solid curve is the prediction of a calculation[19] which takes into account nucleons and the contribution of the pion-exchange current derived from chiral symmetry.

Experiment and theory have made major advances in the three-nucleon problem. The two-body contribution is now theoretically well under control and thus enables us to establish the presence of non-nucleonic processes. Remaining uncertainties are at the level of relativistic effects and of the short-range part of the three-body force.

5.2 The momentum distribution of ^{3}He

In the impulse approximation, the (e,e'p) cross section can be factorized into the elementary electron-nucleon cross section and a spectral function which depends on the energy and momentum distribution of the bound proton. If such a factorization were possible, the momentum distribution of the struck proton could be extracted directly from the experimental data. This factorization, however, is only an approximation and one must take into account final state interactions and meson exchange effects. The experimental data for deuterium are in excellent agreement with theoretical predictions up to momentum of 500 MeV/c.

The electrodisintegration of ^{3}He by means of the (e,e'p) reaction has been studied for momenta up to 600 MeV/c. Figure 7 shows data[24,25] at low momentum. These data are compared to two theoretical predictions. The prediction of Van Meijgaard and Tjon[26] has been obtained by solving the Faddeev equations for the scattering states. The prediction of Laget[28] takes into account final state interactions by a diagrammatic expansion. At higher momentum, final state interactions become important and they have to be taken into account to make a quantitative comparison with experiment. This is shown in figure 8 where all the data taken at Saclay[27] are compared to different predictions[28-31] in the impulse approximation. The general behavior of the distribution is well reproduced, but one would need to include final state interactions to improve the agreement with experimental data. Recent (e,e'p) and (e,e'd) experiments from Nikhef[25] have investigated systematically the effect of correlations, final state interactions and meson exchange currents. Final state interactions are well described by the diagrammatic expansion of Laget[29,33] and a good agreement is observed with theoretical predictions (figures 9 and 10). There is no indication of a significant excess of high momentum components. From these data, one can conclude that up to 600 MeV/c the proton momentum distribution in ^{3}He is reasonably well understood.

However, these data have not really probed the region where short-range correlations are expected to play an important role in nucleon momentum distributions. The present electron accelerators are not able to investigate the very high momentum components where large effects of short range correlations have been predicted. We have to wait for the new generation of multi GeV accelerators to probe the very high momentum contents of nuclei by (e,e'p) experiments. This is one of the exciting challenges for the future.

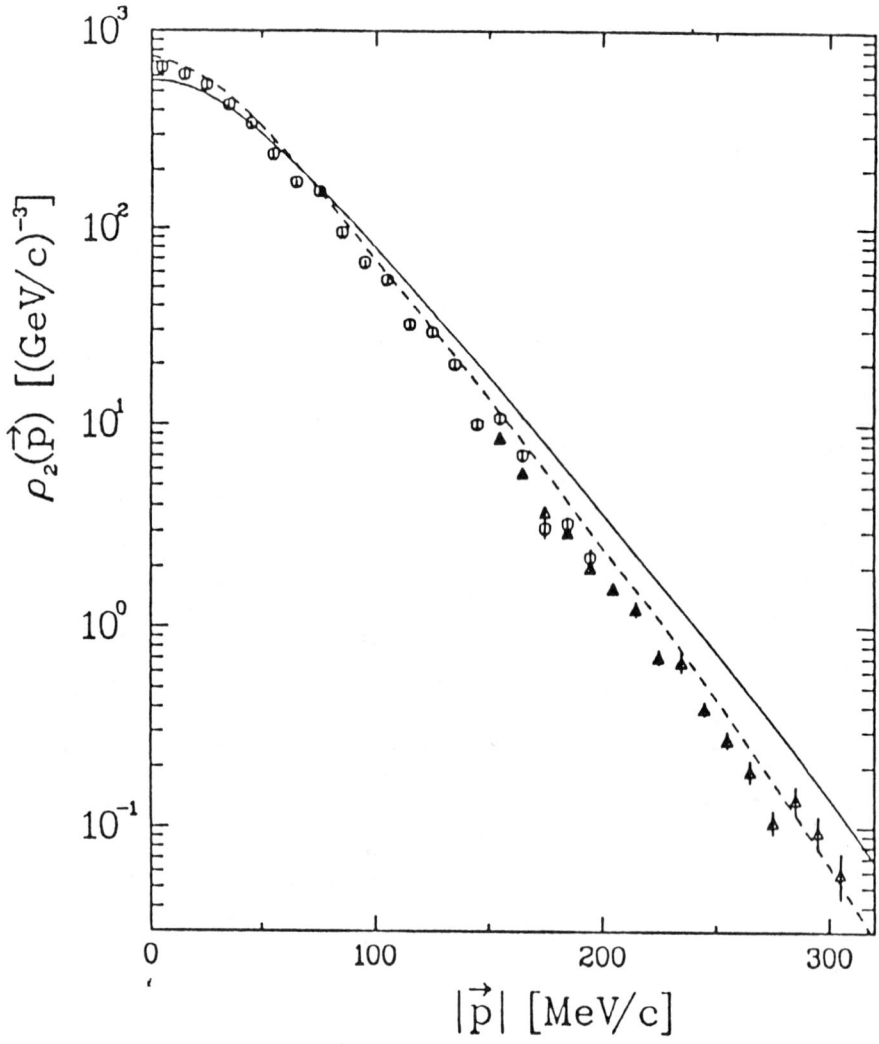

Fig. 7 - Nucleon momentum distribution of ^{3}He for the two-body break up. Solid curve represents the calculation of van Meijgaard and Tjon[26] and the dashed curve shows the result of Laget[29]. The experimental data are from Jans[24] and Keizer[35].

6. TWO-NUCLEON CORRELATIONS

Investigation of two-nucleon correlations requires the coincident detection of two or three reaction products, as in the (e,e'N) or (e,e'2N) channels.

Potentially, two-nucleon knockout experiments (e,e'2N) yield the most direct information on the short range behavior of the NN interaction. Such experiments are today, at the limits of operating facilities. However, recent results from the ongoing (e,e'p) experiments at low duty factor accelerators have provided evidence for the existence of such correlations.

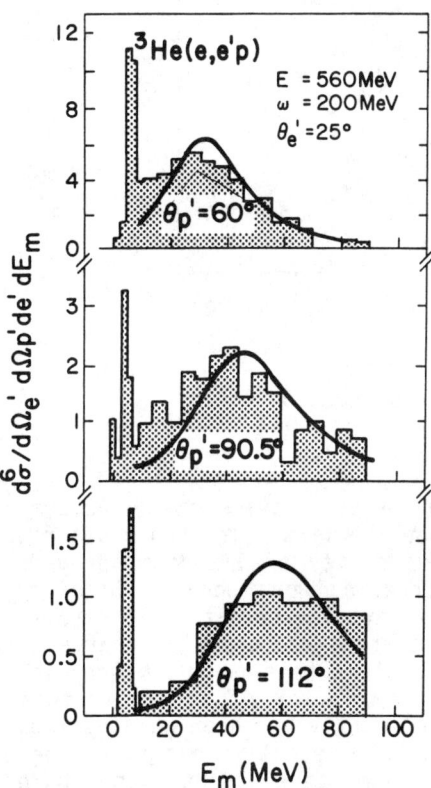

Fig. 11 - Evidence for the existence of correlated pairs in the ^{3}He(e,e'p) spectra[32,33]. The broad peak observed in the three-body break-up channel is the signature of virtual photon absorption on correlated NN pairs. The solid curve is a theoretical prediction of Laget.

The observation of the coupling of the virtual photon to correlated nucleon pairs is found in the study of the three-body break-up observed in the ^{3}He(e,e'p) reaction[32]. Figure 11 shows three missing energy spectra at recoil momentum values of 316 MeV/c, 401 MeV/c and 458 MeV/c respectively. The peak at 5.5 MeV corresponds to the one body knockout reaction to a deuteron final state. A broad peak at higher missing energy values is also evident in the same spectra. This peak which appears at different values of missing energy in each spectrum is due to the break-up of a p-n pair in ^{3}He. It corresponds to the disintegration of a correlated p-n pair at rest. The curves shown in figure 11 are the results of a calculation[33] for ^{3}He based on Faddeev wave functions and which takes into account final state interactions and meson exchange currents. Similar (e,e'p) measurements[34] in ^{12}C find also large increase in the high energy region of the missing mass spectrum.

The ^{3}He(e,e'd)p experiment[35] has allowed the study of two-body mechanisms at low momentum transfers. Cross sections were measured in parallel kinematics with the momentum transfer fixed at q = 380 MeV/c and the recoil momentum varying between 0-200 MeV/c. The experimental coincidence cross section is shown in figure 12. The dotted curve is the expected cross section for the process in which the virtual photon couples to a single nucleon. The full curve is obtained if in addition the coupling to a correlated p-n pair is allowed (in either T=0 or T=1 state).

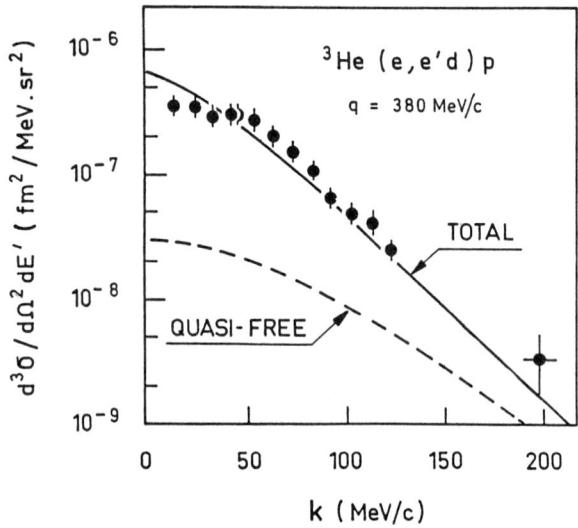

Fig. 12 - ^{3}He(e,e'd) cross section[35,33].

CONCLUSION

Electron scattering experiments have provided very accurate data on the few-nucleon systems. Since the first electron scattering experiment performed here at the University of Illinois, progress has been continuous. After the pioneering work done at Stanford, successive generations of experiments have investigated the structure of few-nucleon systems whith increasing accuracy. The charge and magnetic factors of the trinucleon systems have now been separated up to $Q^2 = 25$ fm^{-2}. The magnetic form factor of the deuteron is known up to 60 fm^{-2}. The scientific community had been waiting these data for nearly 25 years, but it is only recently that they became feasible due to considerable experimental developments. The charge and quadrupole form factors of the deuteron are however still unseparated. Different projects exist which are reviewed by Roy Holt at this conference.

Experimental data cannot be explained in terms of nucleons only. The impulse approximation is not valid even at low momentum transfer. The nucleonic amplitudes which contribute to the cross section have large destructive interferences. This has enabled to isolate the various mesonic processes which occur successively at different momentum transfers. The experimental data are well described by theoretical predictions which take into account meson exchange currents. Up to ~ 15 fm^{-2}, the pion current plays a central role. At higher momentum transfers, the ρ- and Δ meson exchange currents are necessary when the pion-nucleon coupling is assumed to have a finite size. However quite surprisingly with a pointlike coupling to the nucleon, the pion-exchange current is the only one which is needed in addition to the nucleon current to explain magnetic isovector data. This can be easily understood at low momentum transfer, because of chiral symmetry. Since there is no such a constraint at high momentum transfers, the fact that the same observation holds both for the 2 and 3-body systems is quite intriguing. The role of chiral symmetry is also highlighted by the recent SLAC data on the magnetic form factor of the deuteron. Although for a longtime chiral symmetry did not appear to play a role for isoscalar processes, it has recently been shown that isoscalar currents are related to the chiral anomaly. Theoretical predictions in this framework are able to explain the magnetic form factor of the deuteron up to 60 fm^{-2}.

Major progress has been achieved in the understanding of the trinucleon system. Calculational techniques are now reliable. This has enabled to explore the effect of non-nucleonic components and three-body forces. Three-body forces are necessary to explain the 1 MeV missing in the binding energy of ^{3}H and ^{3}He. However, three-body forces are too small to explain the large discrepancies between experiment and theory for the trinucleon charge form factors. Δ-isobar effects are also not sufficient. Recent calculations show that meson-exchange currents are able to reconcile theory and experiment.

Much less is known about momentum distributions and two-nucleon correlations. Though experimental data exist now up to 600 MeV/c, we have not yet reached the region where short-range correlations are expected to play a large role. We are now beginning to reach the limits of the experimental possibilities of the accelerators designed around 1970. New multi-GeV continuous beam accelerators are now the obvious path to the future of investigations in few-nucleon systems. They open a rich perspective of a new class of experiments which is not feasible at present. With the developments now planned at CEBAF very exciting new possibilities will be within our reach. The next 35 years of electron scattering experiments appear full of promises.

ACKNOWLEDGEMENTS

This talk is largely based on a review written in collaboration with C.N. Papanicolas[36]. I am very grateful for his very stimulating and helpful comments. I am also indebted to all my colleagues at Saclay who have contributed to many of the experiments described in this review. Special thanks are due to Mannque Rho for illuminating discussions.

REFERENCES

1. M. Chemtob and M. Rho, Nucl. Phys. A163, 1 (1971).
2. D.O. Riska and G.E. Brown, Phys. Lett. 38B, 193 (1972).
3. M. Rho, Ann. Rev. Nucl. Part. Sci. 34, 531 (1984).
4. R.E. Rand et al., Phys. Rev. Lett. 18, 469 (1967) ;
   D. Ganichot et al., Nucl. Phys. A178, 545 (1972) ;
   G.G. Simon et al., Nucl. Phys. A324, 277 (1979) ;
   M. Bernheim et al., Phys. Rev. Lett. 46, 402 (1981) ;
   S. Auffret et al., Phys. Rev. Lett. 55, 1362 (1985).
5. J.F. Mathiot, Nucl. Phys. A412, 201 (1984).
6. W. Leidemann and H. Arenhövel, Nucl. Phys. A393, 385 (1983).
7. G.E. Brown, M. Rho, W. Weise, Nucl. Phys. A454, 669 51986).
8. A. Buchmann, W. Leidemann, H. Arenhövel, Nucl. Phys. A443, 726 (1985) ;
   D.O. Riska, Physica Scripta 31, 471 (1985).
9. G.E. Brown and M. Rho, Comments Nucl. Part. Phys. 10, 210 (1981).
10. L. Kisslinger, Lect. Notes in Phys. 260, 432 (1986) ;
    Y. Yamauchi, R. Yamamoto, M. Wakamatsu, Phys. Lett. 146B, 153 (1984).
11. S. Auffret et al., Phys. Rev. Lett. 54, 649 (1985) ;
    R. Cramer et al., Z. Phys. C29, 513 (1985) ;
    R. G. Arnold et al., Phys. Rev. Lett. 58, 1723 (1987).
12. M. Gari and H. Hyuga, Nucl. Phys. A264, 409 (1976).
13. P.W. Sitarski, P. Blunden and E.L. Lomon, to be published.
14. R.G. Arnold, C.E. Carlson and F. Gross, Phys. Rev. C21, 1426 (1980) ;
    M.J. Zuilhof and J.A. Tjon, Phys. Rev. C22, 2369 (1980) ;
    F. Coester, private communication.
15. M. Chemtob and S. Furui, Nucl. Phys. A454, 548 (1986).
16. S. Takeuchi and K. Yazaki, Nucl. Phys. A438, 605 (1985).
17. E.N. Nyman and D.O. Riska, Phys. Rev. Lett. 57, 3007 (1986).
18. J.L. Friar, B.F. Gibson and G.L. Payne, Ann. Rev. Nucl. and Part. Sci. 34, 403 (1984) ;
    T. Sasakawa and S. Ishikawa, Few-Body Systems 1, 3 (1986).
19. P. Sauer, Prog. Part. Nucl. 16, 35 (1986) and private communcation ; C. Hajduk, P. Sauer and W. Strueve, Nucl. Phys. A405, 581 (1983) W. Strueve, C. Hajduk and P. Sauer, Nucl. Phys. A405, 620 (1983).
20. H. Collard et al., Phys. Rev. 138, 357 (1965) ;
    P. Dunn et al., Phys. Rev. C27, 71 (1983) ;
    C. Otterman et al., Nucl. Phys. A436, 688 (1985) ;
    J.S. McCarthy, I. Sick and R. Whitney, Phys. Rev. lett. 25, 884 (1970) ; Phys. Rev. C15, 1396 (1977) ;
    J.M. Cavedon et al., Phys. Rev. Lett. 49, 986 (1982) ;
    F.P. Juster et al., Phys. Rev. Lett. 55, 2261 (1985) ;
    D. Beck, Ph.D. Thesis, MIT (1986).
21. J. Carlson, V.R. Pandharipande and R.B. Wiringa, Nucl. Phys. A401, 59 (1983).

22. M. Beyer, D. Drechsel and M.M. Giannini, Phys. Lett. $\underline{122B}$, 1 (1983) ;
    P. Hoodboy and L. Kisslinger, Phys. Lett. $\underline{146B}$, 613 (1984) ;
    J.V. Vary, S.A. Coon and H.J. Pirner, Few-body problems in physics, vol. II, Edited by B. Zeitnitz (North-Holland 1984).
23. E. Hadjimichael, B. Goulard and R. Bornais, Phys. Rev. $\underline{C27}$, 831 (1983) ;
    D.O. Riska, Nucl. Phys. $\underline{A350}$, 227 (1980) ;
    M.A. Maize and Y.E. Kim, Nucl. Phys. A420, 365 (1984).
24. E. Jans et al., Phys. Rev. Lett. $\underline{49}$, $\underline{974}$ (1982).
25. P. de Witt Huberts, Few-body systems, suppl. 1, 373 (1987).
26. E. Van Meijgaard and J.A. Tjon, Few-body systems, suppl. 1, 307 (1987).
27. J. Morgenstern, Few-body systems, suppl. 1, 363 (1987).
28. R. Schiavilla, V.R. Pandharipande and R.B. Wiringa, Nucl. Phys. $\underline{A449}$, 219 (1986).
29. J.M. Laget, Phys. Lett. $\underline{151B}$, 325 (1985) ; Few-body systems, suppl. 1, 271 (1987).
30. H. Meier-Hajduk et al., Nucl. Phys. A395, 332 51985).
31. C. Ciofi degli Atti, E. Pace and G. Salmé, Phys. Lett. $\underline{141B}$, 14 (1984).
32. C. Marchand et al., to be published.
33. J.M. Laget, in New-vistas in electronuclear physics, Eds. E.L. Tomusiak, H.S. Caplan and E.T. Dressler (Plenum, 1986), p. 361.
34. R.W. Lourie et al., Phys. Rev. lett. 56, 2364 (1986).
35. P.W.M. Keizer et al., Phys. Lett. $\underline{157B}$, 255 (1985).
36. B. Frois and C.N. Papanicolas, Ann. Rev. Nucl. Part. Sci. $\underline{37}$ (1987) to be published.

SCALING — FROM MOMENTUM DISTRIBUTIONS TO BOUND NUCLEON PROPERTIES

Ingo Sick

Dept. of Physics, University of Basel, CH-4056 Basel, Switzerland

© American Institute of Physics 1987

SCALING — FROM MOMENTUM DISTRIBUTIONS TO BOUND NUCLEON PROPERTIES

Ingo Sick
Dept. of Physics, University of Basel, CH-4056 Basel, Switzerland

INTRODUCTION

In the development of electron-nucleus scattering over the past 35 years, coherent scattering has played the dominant role, a fact emphasized again by the contributions presented at this meeting. In electron-nucleon scattering, on the other hand, inclusive scattering has provided the most fruitful results. In this talk, I want to emphasise inclusive electron-nucleus scattering, and show that, in analogy with particle physics, its use and interpretation in terms of scaling can provide most valuable results as well.

The word "scaling" stands for a simple fact: the cross section $\sigma(q, \omega)$ for inclusive electron scattering under certain conditions no longer depends separately on the two independent variables momentum transfer q and energy loss ω. It depends on one single variable z which in term depends on q and ω. If such a "scaling" occurs, one can draw important conclusions on the nature of the reaction mechanism, and obtain a measure of the bound-constituent form factor and momentum distribution.

Historically, scaling got into the headlines with the deep inelastic electron-nucleon scattering experiments performed in the late sixties at SLAC. The resulting cross sections, divided by the Mott cross section, were shown to scale in the variable x. From this feature one could draw the conclusions that the nucleon contained massless constituents of fractional charge and pointlike nature, i.e. quarks, and one could measure the quark momentum distributions. During the seventies this scaling feature was systematically exploited in both electron and muon inclusive scattering. With the recent (μ, μ') experiment by the EMC group, x-scaling has again attracted much attention.

In a review paper [1] published in 1975, G. West mentioned another scaling variable, y, which could be used in electron-nucleus scattering. This type of scaling implicitly had been known since a long time as, for the Fermi gas model, the inclusive cross section depends on a fixed combination of q and ω. This y-scaling was ignored, however, and no data suitable for an application were available.

Scaling in terms of a different variable y, modified to make it applicable for the kinematics where scaling might indeed occur, became popular with our data and analysis [2] of inclusive scattering on ^{3}He measured at SLAC up to very large q. There the occurrence of y-scaling was demonstrated for the first time. The benefits and limitations of y-scaling since have been investigated by a number of studies.

Looking to other fields of physics, we realise in retrospect that some of the ideas connected to y-scaling have been exploited for a long time in (keV) electron-atom and (eV) neutron-liquid scattering, although the word "scaling" never appears. Here "infinite" q is so much easier to reach experimentally that the reduction from two to one variables is made from the very beginning in both formalism and analysis of experiment. Several aspects of inclusive scattering have been treated in detail in these applications, and it turns out to be most useful to draw analogies with electron-nucleus scattering at GeV-energies.

2. KINEMATICS

In order to understand how y-scaling comes about, let us consider the kinematics of the scattering process depicted in Fig. 1. In impulse approximation (IA), the energy loss and momentum transfer of the electron are given to one initially bound constituent with an initial total energy of minus the separation energy, and initial momentum k. When neglecting the final state interaction (FSI) of the recoil constituent, energy conservation is expressed by

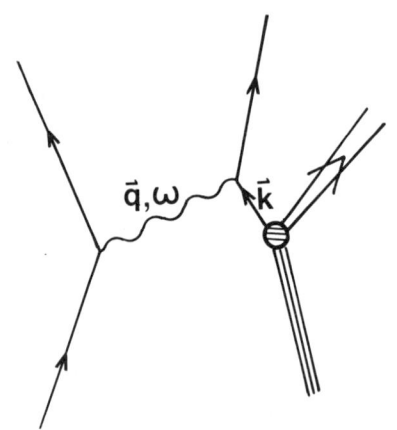

Fig. 1 Impulse Approximation

$$\omega = ((\vec{k} + \vec{q})^2 + m^2)^{1/2} - m + SE \quad (1)$$

In this equation it is useful to split $\vec{k}$ into its components $(k_{\parallel}, k_{\perp})$ parallel and perpendicular to $\vec{q}$. In the limit of very large q, $q \to \infty$, and large ω, terms of order $k_{\perp}$, SE can be neglected, and eq. (1) reads

$$(\omega + m)^2 = K_{\parallel}^2 + 2 K_{\parallel} q + q^2 + m^2 \quad (2)$$

This equation can be solved for $K_{\parallel} = z(q, \omega)$.

Eq. (2) indicates that the variables q, ω no longer are independent. They are linked via the momentum $K_{\parallel}$ of the initially bound constituent. If the cross section is known as a function of one variable, it can be obtained as function of the other one by "scaling" the curve.

The cross section, divided by the elementary electron-constituent cross section at the vertex, σ_{ec}, also can depend on one variable only.

$$\sigma(q, \omega) / \sigma_{ec}(q) \, d\omega = F(z) \, dz \quad (3)$$

The physical meaning of F(z) is obvious; it represents the probability to find a constituent with momentum component z in the initially bound system.

Two cases can be distinguished: For constituent mass zero and pointlike form factor one obtains the scaling variable x occurring in deep inelastic electron-nucleon scattering. For constituents of finite mass and q-dependent form factor one obtains the scaling variable y derived in ref. 2.

$$y \sim (\omega^2 + 2m\omega - q^2) / 2q \qquad (4)$$

In this case, applicable to quasielastic electron-nucleus scattering, y is the momentum component of the nucleon parallel to q.

The above variable y differs in two respects from the one suggested by West. First, the nucleon has a total energy given by minus the separation energy, while West used the Fermi gas prescription + $k^2/2m$. Secondly, the expression defining y is relativistic, while West used nonrelativistic kinematics. Since scaling only occurs at very large q, where the recoil momentum is very large, non-relativistic kinematics are not applicable.

We note that in order to derive scaling, one does not need to deal with the complicated, and often non-transparent, theoretical description of the scattering process as is generally done. Scaling is a consequence of the kinematics and the reaction mechanism only. A simple mechanism, with constrained kinematics, is the central reason for the occurrence of scaling.

3. EXAMPLES

In order to show that experimentally scaling does indeed occur, we display in Fig. 2 the ^{3}He(e, e') data [3] that served to initially establish y-scaling. The inclusive (e, e') data, measured at SLAC at Θ = 8⁰ and energies between 3 and 15 GeV, are shown as a function of q and ω, on a logarithmic scale extending over 7 orders of magnitude. At low q, the quasielastic peak is narrow, and clearly visible. As q increases, the quasielastic cross section rapidly decreases, the peak gets much wider, and the continuum due to nucleon resonance excitation and deep inelastic scattering gets more pronounced.

Plotted as a function of the scaling variable y the data for quasielastic scattering ($\omega < q^2/2m$) is shown again in Fig. 3. The cross sections that in Fig. 2 define a complicated function of q and ω define a unique curve in Fig. 3, depending on y only. The data exhibit a spectacular scaling behaviour.

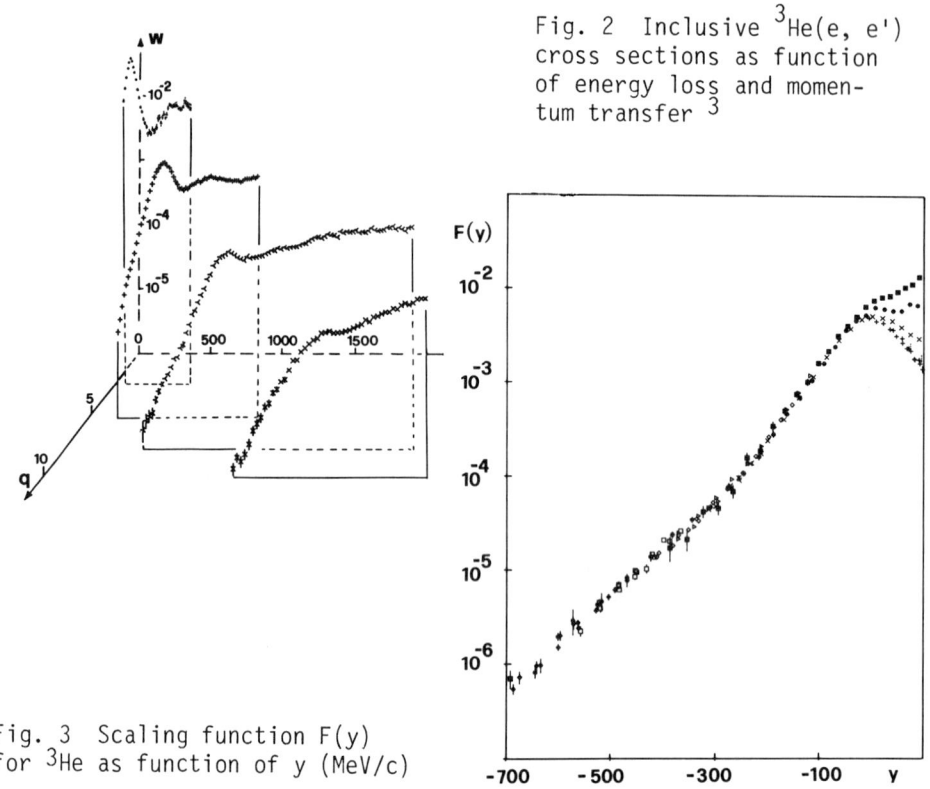

Fig. 2 Inclusive ^{3}He(e, e') cross sections as function of energy loss and momentum transfer

Fig. 3 Scaling function F(y) for ^{3}He as function of y (MeV/c)

4. REACTION MECHANISM

In order to derive scaling, we made two assumptions: the dominance of IA, and the unimportance of FSI. The occurrence of scaling can be used to ascertain the correctness of these assumptions for the kinematical region considered. This is an extremely useful aspect of scaling. For any reaction used to measure nuclear properties in a quantitative way the reaction mechanism must be known first. Scaling provides experimental evidence on it.

As an example, let us discuss the role of meson exchange currents (MEC). At large momentum and energy transfer, we expect the MEC and nucleon resonance excitation dominate over IA quasi-elastic scattering. In the MEC process, the electron transfers q and ω via the exchanged pion to two nucleons, or excites one nucleon to (say) a Δ-resonance which decays via π-exchange with another nucleon. These processes have kinematics that are very different from the one assumed for section 2. No longer are q and ω related simply via eq. (1). The recoil mass basically becomes $2m_N$ rather than m_N. In addition the electron-nucleon vertex form factor is different from the free e-N cross section employed in eq. 3. Accordingly, we no longer expect scaling in terms of y.

To demonstrate this point, we show in Fig. 4 again the data for ^{3}He, this time including the cross sections beyond the maximum of the quasielastic peak, $\omega > q^2/2m$ ($y>0$). In this region the cross sections are dominated by the excitation of the 3/3-resonance. According to the different kinematics and vertex form factor (N-Δ transition), no scaling is expected, a fact borne out by the data.

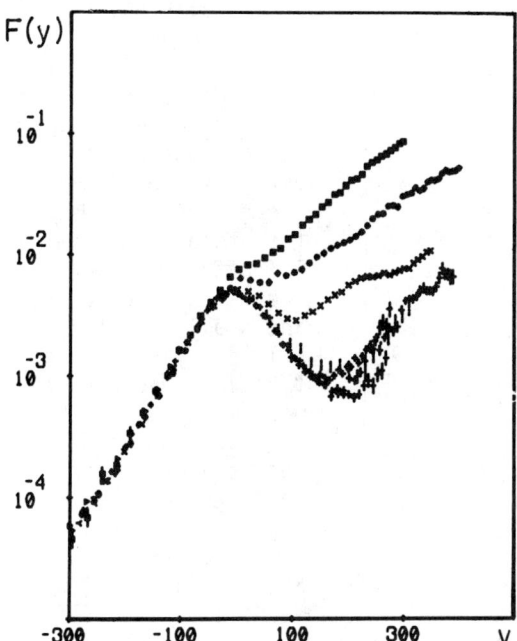

Fig. 4 Scaling function F(y) for ^{3}He including region of Δ-excitation

At large q and very small ω, the IA cross sections become very small due to the smallness of high-momentum components in the nuclear wave function. Under these circumstances one generally has to suspect that MEC become important. The cross sections due to MEC have been calculated for $y<0$ by Donnelly et al.[4] for the kinematical region of the ^{3}He data shown in Fig. 2. Once analyzed in terms of y-scaling, the cross sections show a pronounced non-scaling behaviour (Fig. 5). The important observation is not the one that the MEC cross sections are much smaller than the experimental ones; this might be attributed to a fault of the calculation. The important feature is the non-scaling of non-IA processes, an expected feature which is borne out by the calculation.

Since the experimental data do scale (within a residual non-scaling that can be read from the plots), we can conclude that IA is dominant, and that MEC contributions are smaller than the non-scaling piece of the data (which is small).

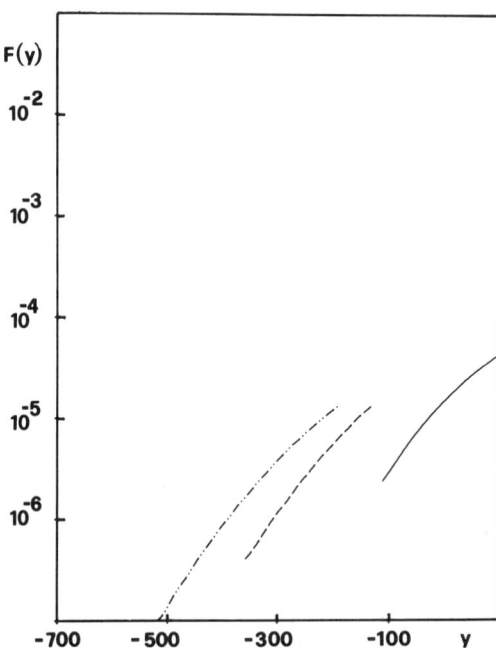

Fig. 5 MEC cross sections analysed in terms of scaling

Similarly, one can show that FSI with the (A-1) nucleus leads to a non-scaling of $F(y)$. The FSI of the knocked-out nucleon depends essentially on its kinetic energy. Neglecting the recoil energy of the (A-1) nucleus and the binding energy, this nucleon energy is given by the electron energy loss, ω. We therefore expect large FSI effects at low ω (recoil nucleon energy) and small FSI at large ω. FSI thus yields an additional dependence of $\sigma(q, \omega)$ on ω, which has no reason to (and does not) scale. Scaling thus can be used to obtain an upper limit on the contribution of FSI to $\sigma(q, \omega)$. (In Fig.3 we have already omitted the region $\omega < 100$ MeV where FSI are expected to be important).

One caveat concerning deductions on the reaction mechanism from scaling should be added. To draw significant conclusions, it is important that the elementary cross section $\sigma_c(q)$ has a large variation over the kinematical range considered. For the data shown in Fig. 2, σ_{ep} changes by a factor of 10^3 for fixed y. Only if $\sigma(q, \omega)$ scales after dividing by a strongly varying function of q (eq. 3) can a significant statement on the reaction mechanism, i.e. dominance of the e-N scattering vertex, be made. If $\sigma_c(q)$ does not vary over a large dynamical range, division by $\sigma_c(q)$ does not change much, and "scaling" (similar in quality before and after division) says nothing on the dominance of the mechanism assumed. The "scaling" in (x, p) reactions with final-state protons of hundreds of MeV in the backward direction, is an illustration for a case where the small variation of the elementary (x, p) cross section allows to draw no conclusion on the reaction mechanism. Evaporation-type spectra "scale" for trivial reasons.

5. MOMENTUM DISTRIBUTIONS

Provided the experimental data scale, F(y) can be used to determine the momentum distribution of the initially bound nucleons.

A priori, a reaction like (e, e' p) seems much more suitable to determine momentum distributions. These can be measured for selected states of the final nucleus, while (e, e') sums over all states. The great interest in F(y) becomes clear as soon as one looks at the y-scale (Fig. 3). Very large values of y can be reached. This is unique, for two reasons:
- With present-day facilities, it is not possible to reach via (e, e' p) these high momenta, where the momentum space density is very small.
- When measuring very small amplitudes in (e, e' p) one will run into difficulties with the reaction mechanism, the FSI of the knocked-out nucleon in particular. Inclusive processes are much less sensitive to FSI; in the limit of closure (a complete set of states available to the recoil) FSI becomes unimportant.

Once one can measure large momenta, one has access to nuclear properties that have been sought for a long time, short-range nucleon-nucleon correlations in particular. Previous attempts to measure large -k components always have failed due to the presence of multi-step processes for reactions involving hadrons. Inclusive electron scattering does not involve the detection of hadrons, and the fate of the recoil-nucleon is of no concern. This is true provided that the length scales relevant for the electron ($1/q$) and the wave length of the recoil nucleon, are small compared to nuclear size and the inter-nucleon distance.

Fig. 3 shows that by (e, e') we can reach values of $K_{||} = y \sim 700$ MeV/c, which corresponds to values of k of $\sim \sqrt{2} \cdot 700$ MeV/c = 1 GeV/c, i.e. very large momenta indeed.

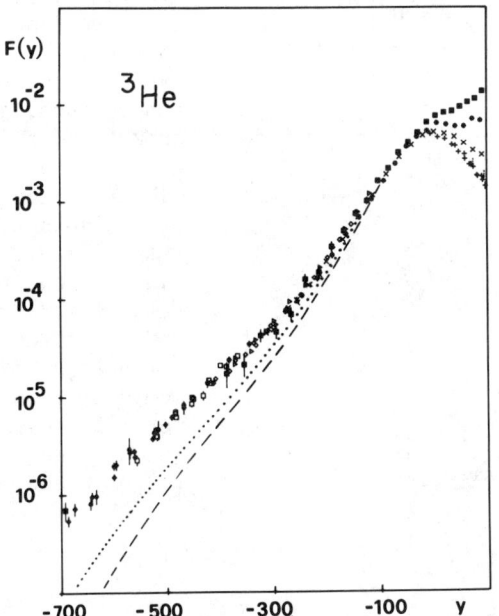

Fig. 6 Experimental data on F(y) for ^{3}He, together with prediction from Faddeev calculations without (dashed) and with (dotted) 3-body force.

In Fig. 6 we compare F(y) for ^{3}He to predictions obtained from Faddeev calculations [5] for the 3-nucleon system performed with the RSC nucleon-nucleon interaction. At small momenta agreement is satisfactory, at very large momenta the Faddeev predictions are too low. The introduction of the Tucson-Melbourne three-body force helps[6], but not enough.

The difference between experiment and calculation can be due to two reasons: At these large momenta, a non-relativistic description can be expected to fail. At the short range one studies at large k, degrees of freedom other than the nucleonic ones considered may play a role; the polarisation of the nucleon distribution by these non-nucleonic wave function components can lead to appreciable changes of $\rho(k)$.

Of particular interest is the momentum space density at $k > k_F$ of nuclear matter. For this observable we today have sophisticated calculations available. In order to measure F(y) for nuclear matter, we recently have carried out an (e, e') experiment at SLAC for A = 4, 12, 27, 56, 197. From this set of measurements [7], taken between 2 and 12 fm^{-1} momentum transfer, one can extrapolate to A = ∞, i.e. eliminate the contribution of the nuclear surface.

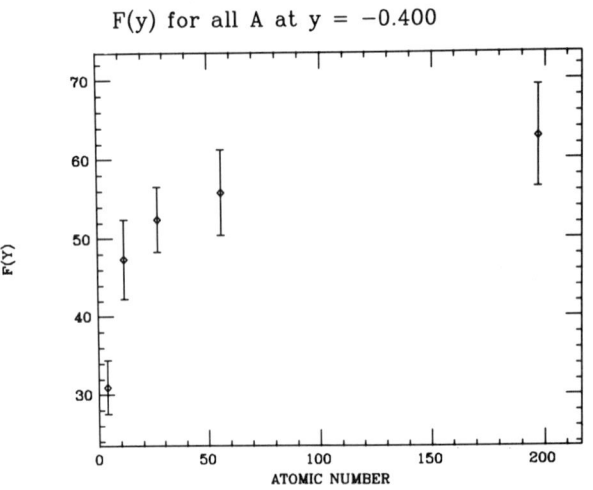

In Fig. 7 we show F(y) at fixed y as a function of mass number. One clearly observes convergence, and an extrapolation of the data plotted against $A^{-1/3}$ allows to reach A = ∞. Fig. 8 shows the data at fixed y and A plotted as a function of momentum transfer. It shows that the data do converge indeed.

Fig. 7 Convergence of F(y) as function of mass number

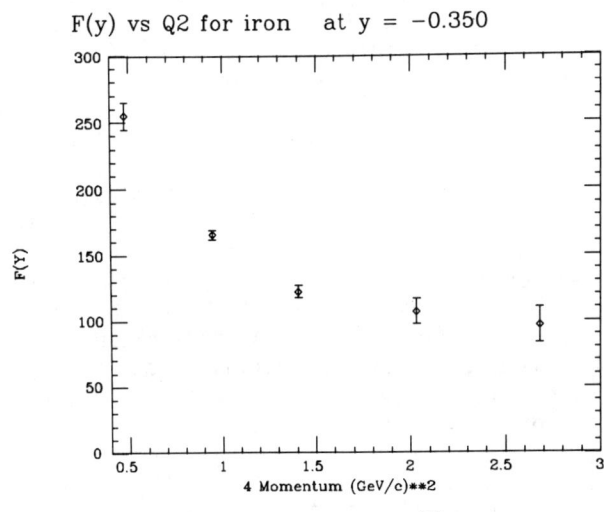

Fig. 8 Convergence of F(y) for ^{56}Fe

Fig. 8 emphasises that scaling is a property expected to occur only in the limit $q \to \infty$, i.e. $q \gg k_F$. The applications to lower-q data ($q \sim 2k_F$, $q^2 \sim 0.25$) often tried cannot lead to significant results.

Since the data analysis of the above experiment is not yet quite complete, I do not want to compare the results to predictions for nuclear matter. It is clear, however, that the inclusive data carry great promise in determining a most interesting property of nuclear matter.

6. MEDIUM EFFECTS ON NUCLEON SIZE

In order to obtain F(y) from $\sigma(q, \omega)$, one divides by the electron-nucleon cross section (eq. 3). If this σ_{eN} does not have the correct q-dependence, the data will not scale. Accordingly, one can use the scaling property to determine the q-dependence of the form factor of the nucleon. This application of scaling is analogous to the one in x-scaling which served to measure convincingly the q-independence of the form factor of quarks, and establish hereby their pointlike nature.

The form factor of the free nucleon is well known from electron-nucleon scattering. The interest in the use of scaling lies in the possibility to determine the q-dependence of the form factor of a nucleon in the medium [8]. Nuclei are dense objects, with an average N-N distance that exceeds the nucleon diameter by $\sim$ 20 % only. Medium effects can be expected to change the bound nucleon, its shape and form factor.

For three reasons, one particular change, an increase in the overall nucleon radius, recently has received extensive attention:
- the EMC effect shows that the probability to find quarks of large momentum ($x \sim 0.7$) is reduced if nucleons are bound in nuclei [9]. Using the uncertainty principle, this indicates an increase in the size of the confinement volume. A general radius increase of the bound nucleon is one possibility to account for this.

- The integral over the longitudinal response function measured via quasielastic electron-nucleus scattering [10, 11] is less than the sum-rule value $Z \cdot \sigma_{ep}$. This could be interpreted as a reduction in $\sigma_{ep}(q)$, indicating an increase in size of the bound proton.
- The q-dependence of $\sigma_{ep}(q)$ measured via (e, e'p) for a given nuclear level and fixed final-state nucleon energy (to minimise the effects of FSI) does not quite correspond to the one expected for the free proton [12]. Again a change of the proton size due to binding could be assumed.

The above pieces of evidence suggest an increase of the radius of ~ 20 %. Unfortunately, the experimental evidence is far from conclusive, and more plausible interpretations exist. Many explanations of the EMC effect have been advanced. The longitudinal sum rule is valid only at $q \gg 2\,k_F$, while data are available at $q \leq 2k_F$; at these transfers Pauli blocking, short-range NN correlations and FSI effects lead to a reduction of the sum rule. In addition, the experimental spectra show a somewhat unphysical shape at large ω, indicating presumably difficulties with radiative corrections at low scattered-electron energy. The small effects observed in (e, e'p) are of a size that could be due to (e, e'n) plus successive (n, p) charge exchange.

Quasielastic scattering analyzed in terms of y-scaling provides a sensitive and less ambiguous means to measure the q-dependence of the bound-nucleon form factor. The momentum transfer is very large ($q \gg 2k_F$), so that FSI, Pauli blocking etc. are small, the corrections due to radiative effects are minor, and no hadrons in the final state are observed.

When using in eq. 3 a $\sigma_{ep}(q)$ of the appropriate q-dependence, one obtains the scaling behaviour one must expect at large q. If one divides by a σ_{ep} of the inappropriate q-dependence, resulting from an incorrect assumption on the bound-nucleon size, scaling is destroyed. The best scaling then can be used to determine the correct q-dependence of the bound-nucleon form factor (size) [8].

In order to extract this information from the (ee') data, we use for the nucleon form factor the dipole expression, which depends on one parameter, the overall size. To judge the quality of scaling, we fit $F(y)$ with a flexible parametrization. The χ^2 of the resulting fit is an indication of the quality of scaling; low χ^2 shows that the different data sets at different q^2 define a unique curve.

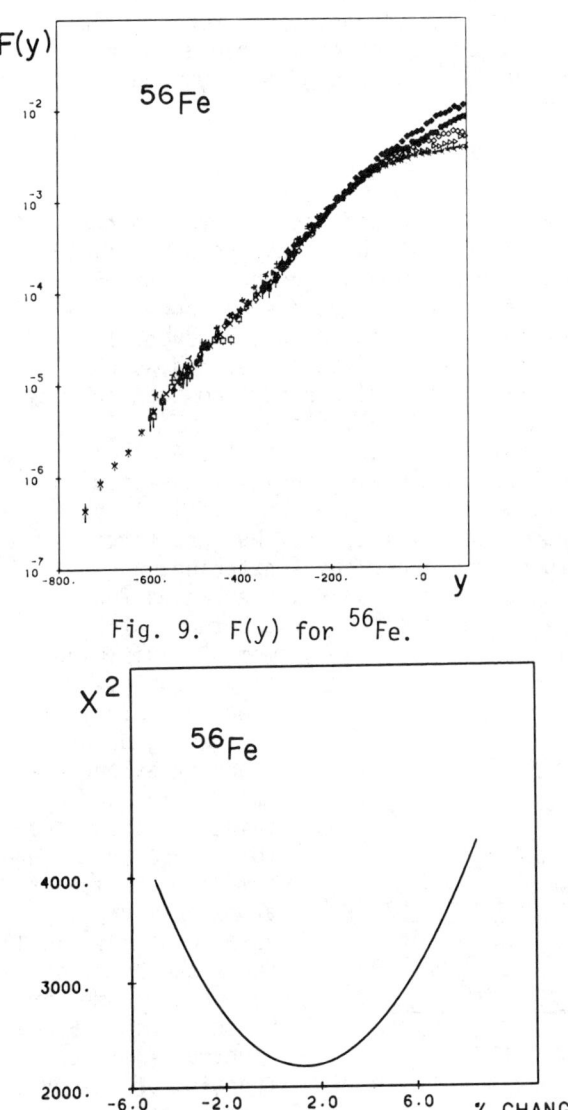

Fig. 9. F(y) for ^{56}Fe.

Fig. 10 Chisquared of fit to F(y) for ^{56}Fe as function of assumed change in nucleon radius

In Fig. 9 we show the Fe (e, e') data of our recent SLAC experiment ($q = 2-12$ fm^{-1}), in Fig. 10 we show the χ^2 as a function of the assumed bound-nucleon size. The difference to the free-nucleon size is 1-2 % in radius. Consideration of various errors leads to an estimate of $\leq$ 3 % radius change due to binding.

This radius change (which refers to a mixture of charge and magnetic properties as the cross sections contain contributions from both) is much smaller than the changes suggested by the pieces of experimental evidence mentioned above. The change found also is much smaller than the ones predicted by some of the theoretical models used to study binding effects [13-15].

The change found does, however, agree with the most general theoretical consideration on binding effects. Amado et al.[16] have shown that a quantum mechanical system, bound by a rather general Hamiltonian, will decrease in size if this system is brought into an attractive potential. The size change depends on the depth V of the potential, and on the energy of the monopole resonance of the system. Using V = 25 MeV and the energy of the Roper resonance yields a $\Delta R/R$ of $\sim$ 2 %.

7. SCALING IN OTHER FIELDS

While y-scaling was first proposed and observed in refs.[1,2] the phenomenon is not really new. If one looks to other fields of physics, the equivalent phenomenon has been exploited for a long time to measure momentum distributions of bound constituents. The name "scaling" does not appear, however. In these applications the limit $q \to \infty$ is easily reached, such that both theory and experiment are discussed from the very beginning in terms of the asymptotic variable y.

It is of interest to make the analogy with these other fields because some of the questions concerning inclusive scattering and scaling there have been investigated in great detail. In particular the role of FSI has been studied, and the results can profitably be applied to electron-nucleus scattering.

Inclusive scattering of electrons of keV energies from atoms has been applied systematically for the determination of momentum distributions of atomic electrons. Inclusive scattering of neutrons of eV energies from solids and liquids has been used to measure the momentum distribution of atoms in matter. As an example, Fig.11 shows the Bethe surface of CO_2, measured by scattering 35 keV electrons [17], as a function of energy loss (atomic units) and momentum transfer k^2. The evolution of the quasielastic

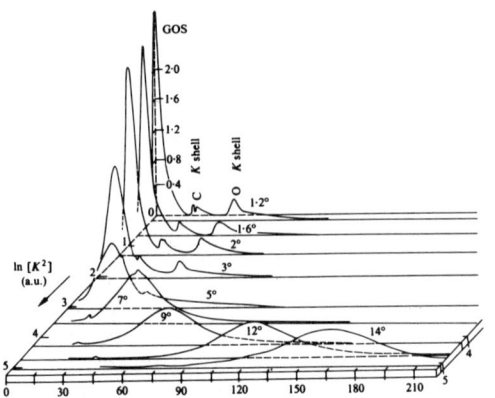

Fig. 11 Bethe surface of CO_2 as function of energy loss and momentum transfer [17].

peak is analogous to Fig. 2 (the elementary nature of the electron leads to the absence of the deep inelastic continuum).

One case I want to discuss in more detail is neutron scattering from superfluid ^{4}He. For this liquid one expects the presence of a certain fraction of atoms with momentum zero, due to the existence of the Bose condensate. In the momentum distribution

$F(y)$, this should yield a $\delta(o)$-like spike superimposed on the normal thermal momentum density.

Experiments on n-^{4}He scattering [18] fail to show this $\delta(o)$ function, although the peak of $F(y \simeq o)$ does have a certain pointedness to it not expected for a purely thermal distribution (Fig. 12). This absence of a δ-function has lead to extensive studies of the effects of the He-He final state interaction on $F(y)$. This atomic interaction is very strong, has the typical Leonard-Jones radial dependence, $r^{-12} - r^{-6}$, and is very repulsive at small inter-Helium distance r (features reminiscent of the NN potential). Calculations [19] of the FSI effects yield two results:

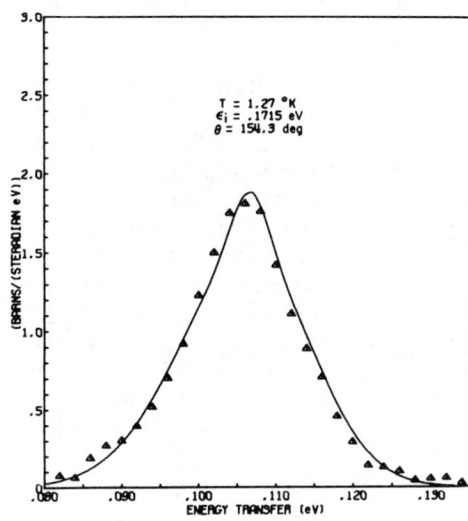

Fig. 12 Response function for neutron-Helium scattering [18]

- The peak of $F(y)$ is shifted, by an amount proportional to the ^{4}He density and the 0^0 scattering amplitude. This shift corresponds to the familiar shift of the quasielastic peak due to nucleon binding.
- The distribution $F(y)$ is folded with a function of a width that depends on Helium density and the total He-He cross section. It is this folding that smears out the δ-function mentioned above.

In liquid Helium, the FSI effects are very large, and they can be used to test whether the theoretical description is correct. Moreover, the He-He total cross section shows minima and maxima as a function of collision energy. The folding therefore is done with a width that depends on momentum transfer, and this leads to a q-dependent width of the quasielastic peak and $F(y)$. This oscillatory q-dependence is indeed observed, Fig. 13, and agrees with predictions.

The FSI effects deduced in connection with neutron scattering can be translated to the case of interest here, nuclear physics. For recoil nucleon energies of > 100 MeV (i.e. energy loss $\omega - \omega_{el} \widetilde{>}$ 100 MeV) the FSI effects, in particular the folding width, is of orden of the bin-size of the experimental data, and can be

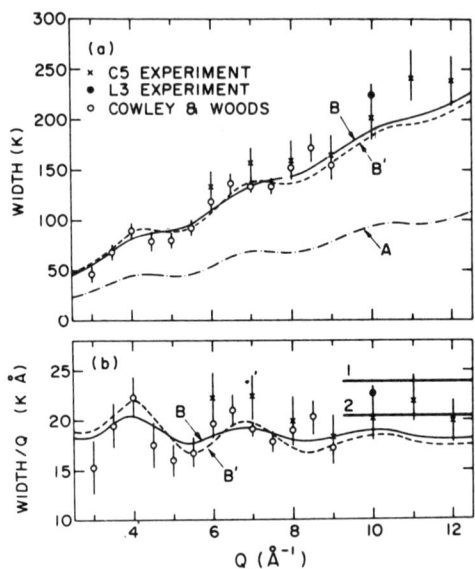

Fig. 13 Width of n-^{4}He quasielastic peak as function of momentum transfer

neglected. For low energy loss, large FSI effects can be expected (and are observed). Accordingly, we have shown in the figures of F(y) only data with $\omega - \omega_{el} > 100$ MeV.

8. CONCLUSION

In this talk I have discussed a number of aspects of y-scaling, pointing out its great benefits and some of its limitations. Clearly, this is not a closed subject. Approximations like the neglect of $K_{\perp}$ can largely be eliminated. For the study of nucleons of very large momentum, the role of the energy of the nucleon which initially is far off-shell needs further investigation.

Y-scaling provides a rich source of information on nuclei, in analogy with the application of x-scaling to the nucleon. From the scaling property of the data, experimental knowledge on the reaction mechanism can be obtained, a prerequisite for quantitative application. The scaling function at large y allows to measure the nucleon momentum distribution at large k, a quantity long sought for in the search for effects of NN short range correlations. In addition, accurate information on the form factor and size of the bound nucleon can be extracted.

For the exploitation of scaling, one needs data at very large momentum transfer; only there the approximations made become valid, and a quantitative interpretation be performed. Such data can be (marginally) measured with today's facilities, and much improved upon with the facilities that will be available at CEBAF.

REFERENCES

1. G.B. West, Phys. Rep. 18 C (1975) 264
2. I. Sick, D. Day, J.S. McCarthy, Phys. Rev. Lett. 45 (1980) 871
3. D.Day et al., Phys. Rev. Lett. 43 (1979) 1143

4. T.W. Donnelly et al., Phys. Lett. 76 B (1978) 383 and priv.com.
5. R.A. Brandenburg et al., Phys. Rev. C 12 (1975) 1368
6. T. Sasakawa, Lect. Notes in Phys. 260 (1986) 150
7. D. Day et al., to be published
8. I. Sick, Phys. Lett. 157 B (1985) 13
9. J.J. Aubert et al., Phys. Lett. 123 B (1983) 123
10. R. Altemus et al., Phys. Rev. Lett. 44 (1980) 965
11. Z.E. Meziani et al., Phys. Rev. Lett. 52 (1984) 2130
12. G. van Steenhoven et al., Phys. Rev. Lett. 57 (1986) 182
13. L.S. Celenza, A. Rosenthal, C.M. Shakin, Phys. Rev. Lett. 53 (1985) 892
14. M.Oka, Phys. Lett. 165 B (1985) 1
15. M. Ericson, M. Rosa-Clot, preprint CERN-TM-4420/86
16. R. Amado et al., to be published
17. A.L. Bennani et al., Chem. Phys. Lett. 41 (1976) 2170
18. P. Martel et al., Journ. Low Temp. Phys. 23 (1976) 285
19. L.J. Rodriguez, H.A. Gersch, H.A. Mook, Phys. Rev. A9 (1974) 2085

ELECTRON SCATTERING AND CORRELATIONS IN NUCLEI

V. R. Pandharipande
University of Illinois at Urbana-Champaign
1110 West Green Street, Urbana, IL

ABSTRACT

We briefly review mean-field and microscopic theories of nuclei. The main emphasis is on correlations that constitute the difference between the mean-field and microscopic descriptions of nuclear ground states. These correlations influence the two-particle distribution function, and we discuss its measurement from the longitudinal response function in electron scattering. It appears that the latest measurements of the longitudinal structure function in ^{3}He nucleus by the Saclay group indicate presence of correlations, but are not accurate enough to draw interesting conclusions. The situation in heavier nuclei is also discussed. The correlations also influence the one-particle density matrix $\rho_1(\vec{r},\vec{r}')$. In correlated systems the $\rho_1(\vec{r},\vec{r}')$ can be expressed as a sum of contributions from fractionally occupied single particle natural orbits. We discuss the natural orbits of a Bose atomic liquid ^{4}He helium drop for illustration, and review the measurements of occupation numbers in nuclei by electron scattering.

I. MEAN-FIELD AND MICROSCOPIC THEORIES

The mean-field theory[1] has provided an excellent starting point to study a variety of nuclear phenomenon. In this theory the wave function of a closed shell nucleus is assumed to be an antisymmetrized product of single particle wave functions $\phi_i(x)$:

$$\Phi(x_1 \ldots x_A) = A \prod_{i=1,A} \phi_i(x_i). \qquad (1.1)$$

The x_i include positron $\vec{r}_i$, spin $\vec{\sigma}_i$ and isospin $\vec{\tau}_i$, however for brevity we often ignore spin and isospin degrees of freedom and use $\vec{r}_i$ in place of x_i. The one-particle density distribution $\rho(r)$ is obtained as:

$$\rho(\vec{r}) = \sum_{i=1,A} \phi_i^2(\vec{r}), \qquad (1.2)$$

© American Institute of Physics 1987

and in the simplest versions of the mean-field theory, called density-dependent Hartree-Fock[1] or energy-density functional theory[2], the single particle potential V(r) is a prescribed functional of $\phi_i(\vec{r})$. The single particle Schrödinger equation:

$$-\frac{\hbar^2}{2m} \nabla_i^2 \phi_i(\vec{r}) + V\{\phi_{j=1,A}(\vec{r}')\} \phi_i(\vec{r}) = \varepsilon_i \phi_i(\vec{r}), \qquad (1.3)$$

is solved consistently. The observed nuclear charge distributions[3] are reasonably, though not exactly, explained with this theory[4] using fairly simple functionals $V\{\phi_{i=1,A}(\vec{r})\}$. However, since there exists a wave function Φ that will reproduce a given $\rho(r)$, any difference between theoretical and experimental charge distribution can always be attributed to shortcomings of the assumed functional $V\{\phi_{i=1,A}(\vec{r})\}$.

In contrast the microscopic nuclear theory is based on the Hamiltonian:

$$H = \sum_{i=1,A} \frac{\hbar^2}{2m} \nabla_i^2 + \sum_{i<j \leq A} v_{ij} + \sum_{i<j<k \leq A} V_{ijk}. \qquad (1.4)$$

There are many models of the two-nucleon interaction v_{ij}, they all include one-pion-exchange potential and fit the nucleon-nucleon scattering data at laboratory energies of < 400 MeV. The models of the three-nucleon interaction[5] include two-pion-exchange potential, and some attempt to fit the triton energy. It appears that it is necessary to include the three-nucleon interaction in the nuclear Hamiltonian to obtain reasonable properties of light nuclei and nuclear matter.[5,6]

In this theory nuclear wave functions are obtained by solving the many-body Schrödinger equation:

$$H\Psi_n(x_1....x_A) = E_n \Psi_n(x_1....x_A). \qquad (1.5)$$

We can only calculate the ground state of the A = 3 nuclei ^{3}H and ^{3}He exactly by solving the Faddeev equations.[7] Approximate, and possibly quite accurate, solutions also exist for A = 4 ^{4}He nuclei and nuclear matter with the variational method[8], and for nuclear matter with the Brueckner-Bethe-Goldstone method.[9] It is difficult to solve the Schrödinger Eq. (1.5) for nuclei with $10 < A < \infty$ due to the strong spin, isospin and tensor forces between nucleons. However, attempts have been made to calculate the ground state of ^{16}O and ^{40}Ca nuclei.[10,11]

The variational wave functions of A = 3 and 4 nuclei and nuclear matter[8] have the form:

$$\Psi_o(^3\text{He}) = (S F_{12} F_{23} F_{31}) A(\uparrow p \downarrow p \uparrow n), \qquad (1.6)$$

$$\Psi_o(^4\text{He}) = (S \prod_{i<j\leq 4} F_{ij}) A(\uparrow p \downarrow p \uparrow n \downarrow n), \qquad (1.7)$$

$$\Psi_o(\text{NM}) = (S \prod_{i<j} F_{ij}) \Phi_{FG}, \qquad (1.8)$$

where Φ_{FG} is the Fermi-Gas wave function and F_{ij} is correlation operator. We note that the deuteron wave function can also be written in a similar form:

$$\Psi_o(^2\text{H}) = F_{12} A(\uparrow p \uparrow n). \qquad (1.9)$$

The F_{ij} contain central, tensor and other correlations:

$$F_{ij} = f^c(r_{ij}) + f^{t\tau}(r_{ij}) S_{ij} \tau_i \cdot \tau_j + \text{other terms}. \qquad (1.10)$$

The $f^c(r_{ij})$ and $f^{t\tau}(r_{ij})$ are shown in Figs. 1 and 2. The $f^c(r_{ij})$ of all nuclei have a minimum at $r_{ij} = 0$ due to the repulsive core in the nucleon-nucleon interaction. The $f^c(r_{ij})$ of A = 2-4 nuclei goes to zero as $r_{ij} \to \infty$ to reflect the finite size of these nuclei, while $f^c(r_{ij} \to \infty) = 1$ in nuclear matter. The tensor correlation increases in magnitude as we go from A = 2 to 4, but it is smaller in nuclear matter presumably due to Pauli exclusion effects. The large $f^{t\tau}(r_{ij})$ in A = 2, 3 and 4 nuclei are responsible for the observed asymptotic D- to S-wave ratios in their (n+p), (d+p) and (d+d) breakup channels[8,12-14] respectively. The f^c and $f^{t\tau}$ constitute most of F in light nuclei, and they are also the most important correlations in nuclear matter. However, other correlations associated with $\vec{\sigma}_i \cdot \vec{\sigma}_j$, $\vec{\tau}_i \cdot \vec{\tau}_j$ and $\vec{L} \cdot \vec{S}$ operators also need to be considered, especially in nuclear matter.

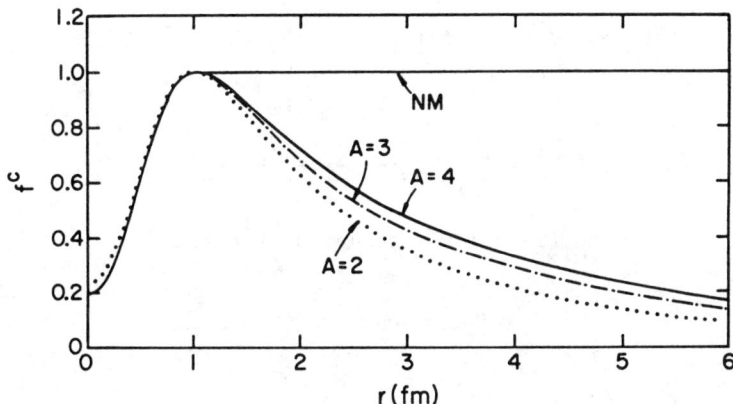

Fig. 1 The central correlation in A = 2,3,4 nuclei and nuclear matter.

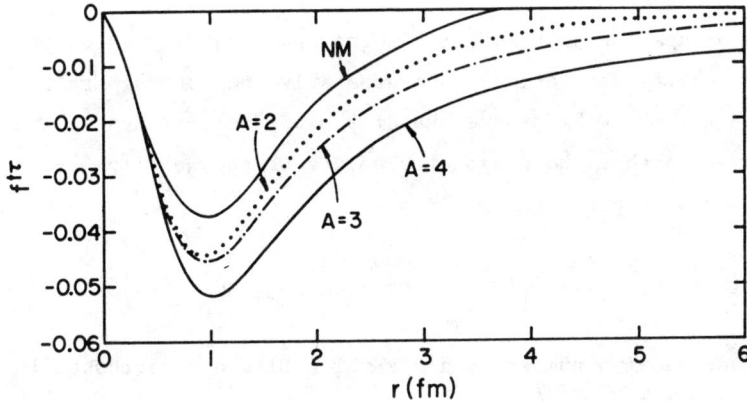

Fig. 2 The $\tau_i \cdot \tau_j S_{ij}$ (tensor) correlation in A = 2,3,4 nuclei and nuclear matter.

In this paper I will discuss the differences between the mean field ground state Φ_0 and the correlated ground state Ψ_0 that can be measured by electron scattering. The two-particle density $\rho_2(\vec{r}_1, \vec{r}_2)$ gives the probability of finding simultaneously a particle at $\vec{r}_1$ and another at $\vec{r}_2$. It is obviously affected by the correlations, and its measurement is discussed in Section II. The one particle density matrix:

$$\rho_1(\vec{r}_1,\vec{r}_2) = \langle 0|a^\dagger(\vec{r}_1)a(\vec{r}_2)|0\rangle, \qquad (1.11)$$

is also influenced by correlations. In the mean field approximation we obtain:

$$\rho_1(\vec{r}_1,\vec{r}_2) = \sum_i n_{MF}(i)\, \phi_i^*(\vec{r}_1)\phi_i(\vec{r}_2), \qquad (1.12)$$

$$n_{MF}(i) = 1 \text{ for occupied states}, \qquad (1.13)$$

$$= 0 \text{ for unoccupied states}, \qquad (1.14)$$

while for the correlated state

$$\rho_1(\vec{r}_1,\vec{r}_2) = \sum_i n(i)\, \psi_i^*(\vec{r}_1)\psi_i(\vec{r}_2). \qquad (1.15)$$

The occupation numbers $n(i)$ in the above equation are not necessarily zero or one, and generally the natural orbits[15] $\psi_i \neq \phi_i$. We will assume that the mean-field states ϕ_i are chosen so that the diagonal elements of the density matrix are correctly reproduced,

$$\rho_1(\vec{r},\vec{r}) = \sum_i n(i)|\psi_i(r)|^2 \equiv \rho(r) = \sum_i n_{MF}(i)|\phi_i(r)|^2. \qquad (1.16)$$

The occupation numbers and natural orbits are discussed in Section III.

The main problem in the microscopic theory of large ($A \sim 100$) nuclei is the treatment of tensor correlations. The ground states of ~ 100 helium atoms (drops of liquid ^{3}He or ^{4}He can be calculated by either variational or Green's function Monte Carlo methods[16-18], because the correlations in these systems do not depend upon the spins. I will often use liquid helium drops to illustrate phenomenon that may also occur in nuclei. The correlations in liquid helium are much stronger than in nuclei, and so their effects are more visible.

II. THE PAIR DISTRIBUTION FUNCTIONS

The pair distribution function $\rho_2(\vec{r}_{12})$ is defined as:

$$\rho_2(\vec{r}_{12}) \equiv \frac{1}{A} \int d^3 R_{12} \rho_2(\vec{r}_1, \vec{r}_2), \tag{2.1}$$

$$\vec{r}_{12} = \vec{r}_1 - \vec{r}_2, \quad R_{12} = \frac{1}{2}(\vec{r}_1 + \vec{r}_2). \tag{2.2}$$

The $\rho_{pp}(\vec{r})$ is the $\rho_2(\vec{r})$ for two protons. The $\rho_{pp}(\vec{r})$ and $\rho_2(\vec{r})$ are normalized so that their volume integrals equal (Z-1) and (A-1) respectively. It is known[19,20] that the longitudinal structure function $S_L(k)$ obtained by integrating the longitudinal response $S_L(k,\omega)$:

$$S_L(k) = \int_{\omega_{el}^+}^{\infty} \frac{S_L(k,\omega) d\omega}{|\tilde{G}_E(k,\omega)|^2} \tag{2.3}$$

can be used to measure $\rho_{pp}(\vec{r})$. In Eq. (2.3) $\tilde{G}_E(k,\omega)$ is the proton form factor, and the lower limit ω_{el}^+ is to exclude elastic scattering. We use the free proton form factor as advocated by Sick.[21] It has been shown that

$$S_L(\vec{k}) = \tilde{\rho}_{pp}(\vec{k}) - Z\, \tilde{F}_{el}^2(\vec{k}) + 1 + \text{corrections}, \tag{2.4}$$

where overhead ~ denote Fourier transforms,

$$\tilde{X}(\vec{k}) = \int d^3 r\, e^{i\vec{k}\cdot\vec{r}}\, X(\vec{r}), \tag{2.5}$$

and $\tilde{F}_{el}(\vec{k})$ is the Form-factor of proton density distribution $\rho_p(\vec{r})$,

$$\tilde{F}_{el}(\vec{k}) = \frac{1}{Z} \int d^3 r\, e^{i\vec{k}\cdot\vec{r}}\, \rho_p(\vec{r}). \tag{2.6}$$

The corrections in Eq. (2.4) come from the scattering of electrons by neutrons and virtual pions. The former have been calculated in Ref. 20 and are small (~ 0.01) for $k < 600$ MeV/c. The pionic exchange current corrections are also expected to be small for longitudinal scattering.

The $\rho_{pp}(r)$, as obtained from microscopic theory[20] using the Argonne[22] v_{14} model of v_{ij} and Urbana[8] model-VII of V_{ijk}, for ^{3}He and ^{4}He nuclei are shown in Fig. 3. The $\rho_{pp}(r)$ have a dip at $r = 0$ attributed to the repulsive core in the nucleon-nucleon interaction. In ^{3}He the $\rho_{pp}(r)$ obtained from variational wave function is in good agreement with that from the more accurate Faddeev wave function.[7] The calculated $S_L(k)$ of ^{3}H, ^{3}He and ^{4}He nuclei are shown in Fig. 4, along with the experimental data of the Saclay group[23] on ^{3}He. We see that there is good agreement between theory and experiment in ^{3}He, however we must examine the effect of the errors in the experimental data.

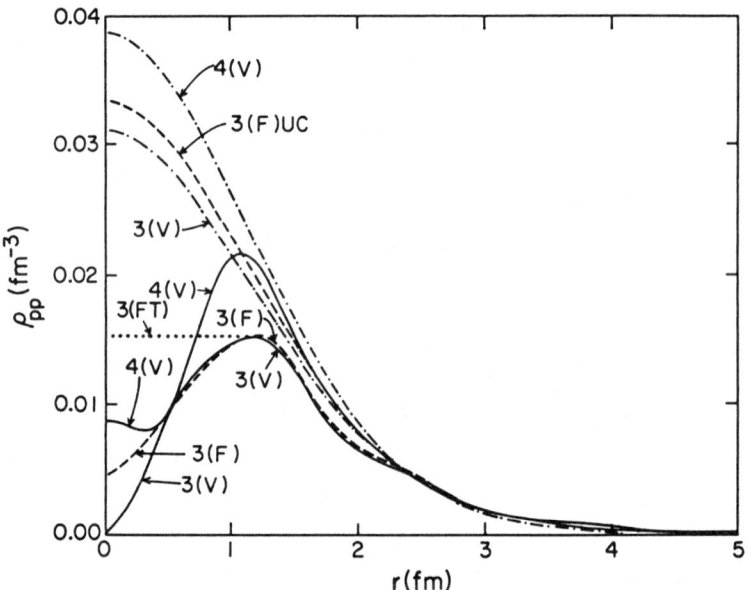

Fig. 3 The $\rho_{pp}(r)$ in ^{3}He and ^{4}He. The full and dash-dot lines give the $\rho_{pp}(r)$ and $\rho_{pp,uc}(r)$ obtained from variational wave functions, while the dashed lines show results obtained with the Faddeev wave function for ^{3}He. The dotted curve labeled 3(FT) is a schematic $\rho_{pp}(r)$ that does not have a hole at $r = 0$.

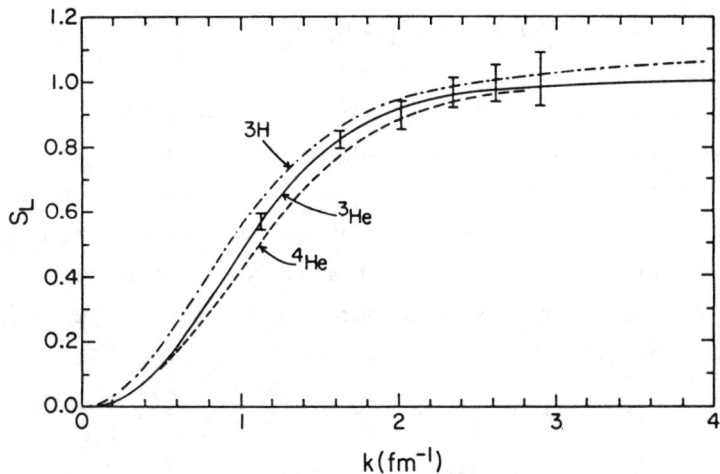

Fig. 4 The $S_L(k)$ of light nuclei. The experimental data points show the $S_L(k)$ of ^{3}He nucleus.

If the motion of the protons is completely uncorrelated the two-proton density $\rho_{pp,uc}(\vec{r}_1,\vec{r}_2)$ is related to the one-proton density $\rho_p(r)$:

$$\rho_{pp,uc}(\vec{r}_1,\vec{r}_2) = \frac{Z-1}{Z} \rho_p(\vec{r}_1)\rho_p(\vec{r}_2). \qquad (2.7)$$

The uncorrelated proton pair distribution function is defined as:

$$\rho_{pp,uc}(\vec{r}_{12}) = \frac{1}{Z}\int d^3R_{12}\, \rho_{pp,uc}(\vec{r}_1,\vec{r}_2), \qquad (2.8)$$

and it is also shown in Fig. 3. The longitudinal structure function $S_{L,uc}(k)$ of a ^{3}He nucleus in which the correlations between the two protons have been artificially switched off is obtained from the $\rho_{pp,uc}(\vec{r}_{12})$,

$$S_{L,uc}(\vec{k}) = \tilde{\rho}_{pp,uc}(\vec{k}) - Z\,\tilde{F}^2_{el}(\vec{k}) + 1$$

$$= 1 - \tilde{F}^2_{el}(\vec{k}). \qquad (2.9)$$

Thus only the difference

$$M(\vec{k}) = 1 - \tilde{F}_{el}^2(\vec{k}) - S_L(\vec{k}) \tag{2.10}$$

provides a measure of correlations. This difference is compared in Fig. 5 with that obtained from the theoretical $\rho_{pp}(\vec{r})$ and a $\rho_{pp}(r)$ that does not have a dip at $r = 0$ but has a flat top as illustrated in Fig. 3. We see that the experimental data cannot differentiate between these two proton pair distribution functions. The accuracy of measurement needs to be improved by a factor of $\sim$ five.

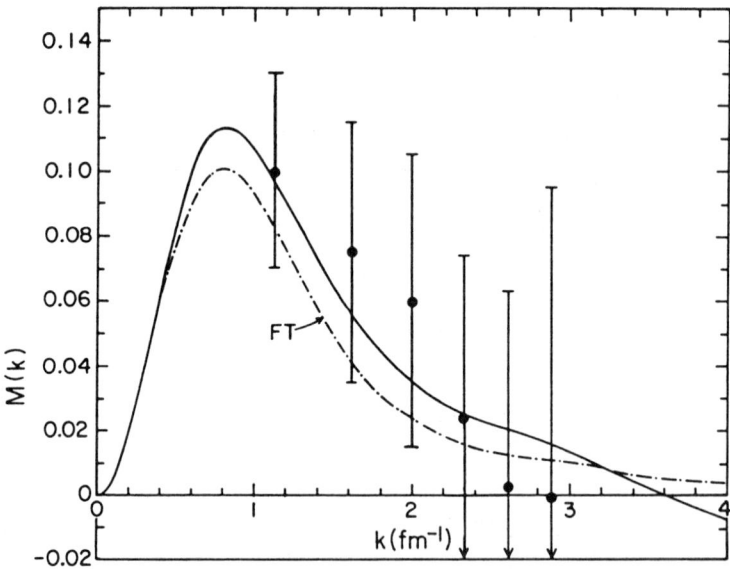

Fig. 5 The M(k) in ^{3}He nucleus.

Even though the $S_L(k,\omega)$ cannot be calculated as accurately as the $S_L(k)$ it is interesting to examine our understanding of it in ^{3}He nucleus. In Fig. 6 we compare results of a recent theoretical calculation[24] of $S_L(k,\omega)$ using correlated final states and the variational ground state with the Saclay data. There is a reasonable agreement between theory and experiment. The theoretical curve is obtained with the form factor

$$\tilde{G}_p^E(k) = \left(1 + \frac{k^2}{4M^2}\right)^{-1/2} \left(1 + \frac{k^2}{\Lambda^2}\right)^{-2}, \qquad (2.11)$$

with $\Lambda = 855$ MeV.

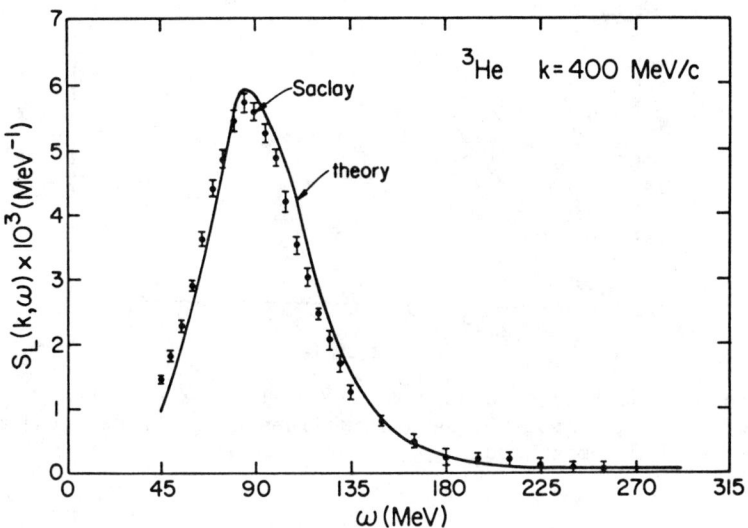

Fig. 6   The theoretical and experimental $S_L(k,\omega)$ of ^{3}He nucleus at k = 400 MeV/c.

The theoretical and experimental studies of $\rho_{pp}(r)$ in heavier nuclei are in much poorer shape. We[20] have calculated the structure functions $S(k)$ of many drops of helium liquids to study their dependence on the number of particles A. The results of these calculations, shown in Figs. 7 and 8, indicate that the $S(k)$ is a smooth function of A approaching that of the liquid as A increases.

The $\rho_2(r)$ and $\rho_{pp}(r)$ in nuclear matter can be easily calculated using variational wave functions, and they are shown in Fig. 9 along with those for the mean field Fermi-gas wave function. The calculated $S_L(k)$ of nuclear matter is not too different from that of the ^{4}He nucleus, and thus, in first approximation we may compare the $S_L(k)$ of nuclei having A > 4 with that of nuclear matter (Fig. 10). The experimental data

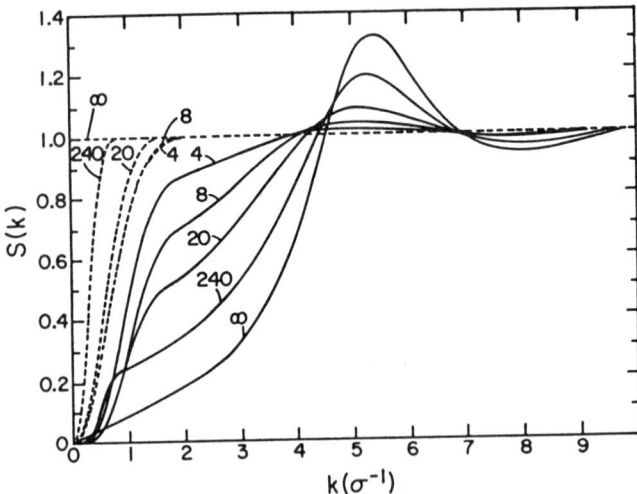

Fig. 7 The S(k) of atomic Bose ^{4}He liquid drops having 4, 8, 20, 240 and ∞ atoms. The dashed lines give the uncorrelated $S_{uc}(k)$ for comparison.

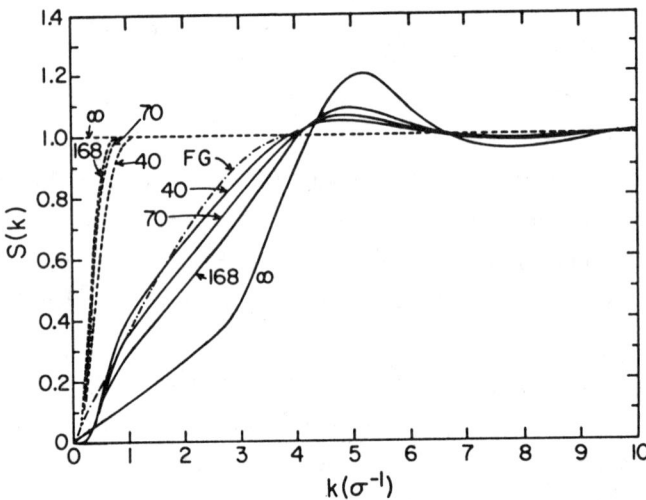

Fig. 8 The S(k) of atomic Fermi ^{3}He liquid drops having 40, 70, 168 and ∞ atoms. The dashed lines give the uncorrelated $S_{uc}(k)$ and the dash-dot line gives the Fermi-gas $S_{FG}(k)$.

compared with experimental data from Refs. 26 and 28 in Fig. 11, and again there is a reasonable agreement between theory and experiment. Approximately 10% of the strength is in a tail extending beyond ω = 240 MeV. The data points shown in Fig. 10 are obtained by integrating the $S_L(k,\omega)$ only inside the quasi-elastic peak, and thus they should not be compared with the $S_L(k)$, but with the dashed curve in Fig. 10 that shows the strength in the quasi-free peak. Unfortunately the calcium data points are below this curve also. The problem is that the experimental response at large ω is smaller than the calculated (Fig. 11). There is good agreement between the observed and calculated response at small ω at all values of k, indicating that a change in the proton form factor[29] may not be the cause of the apparently small observed $S_L(k)$. Clearly more theoretical and experimental studies are needed to understand the $S_L(k)$ of heavy nuclei.

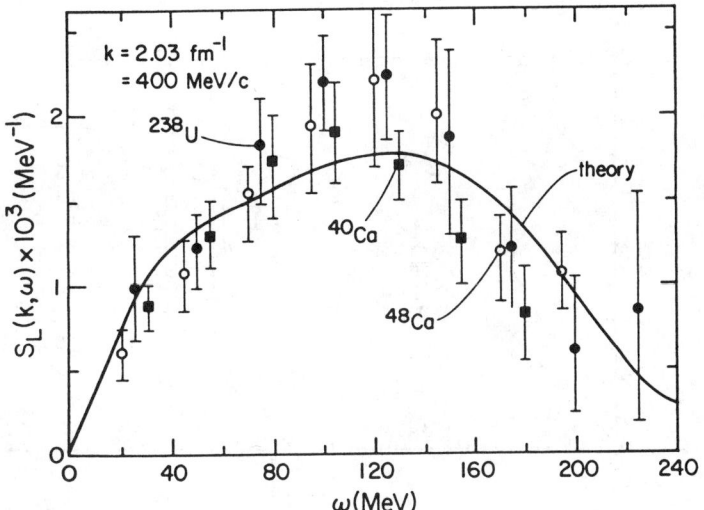

Fig. 11 The observed $S_L(k,\omega)$ of ^{40}Ca, ^{48}Ca and ^{238}U at k = 400 MeV/c is compared with theoretical estimate.

shown in Fig. 10 is from Refs. 25 and 26, and are generally smaller than the theoretical prediction.

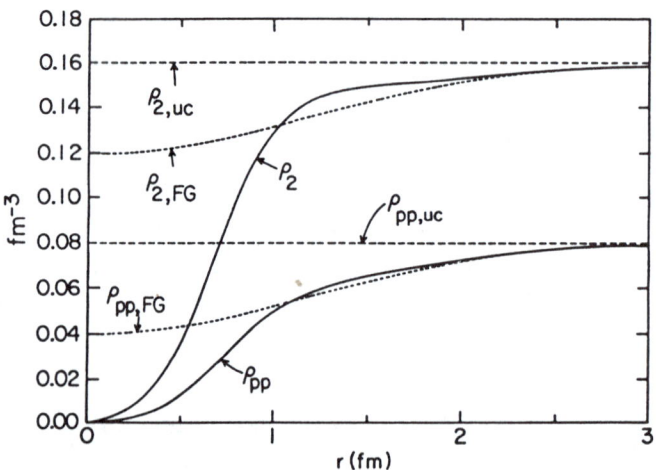

Fig. 9 The correlated (full lines), Fermi-gas (short dashed lines) and uncorrelated (long dashed lines) distribution functions in nuclear matter.

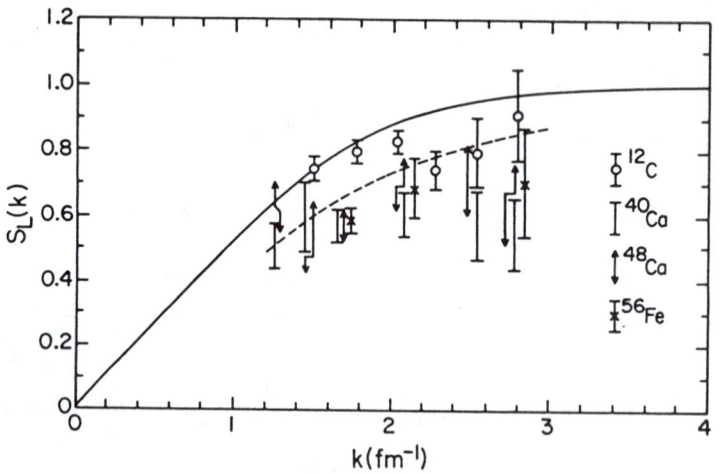

Fig. 10 The $S_L(k)$ of nuclear matter (full line) is compa with data on various nuclei. The dashed line shows the strength in the quasi-free peak only, and it is more reasonable to compare the experimental data with it.

The $S_L(k,\omega)$ in nuclear matter has been recently esti using correlated states. The results at k = 400 MeV/c a

III. NATURAL ORBITS AND OCCUPATION NUMBERS

Pieper et al[30] have attempted to calculate the natural orbits in helium liquid drops using variational wave functions. They calculate the elements ρ_{ij} of the one-particle density matrix using a complete set of functions $\xi(\vec{r})$:

$$\rho_{ij} = \int \xi_i(\vec{r}_1) \rho(\vec{r}_1, \vec{r}_1') \xi_j^*(\vec{r}_1') d^3 r_1 d^3 r_1'$$

$$= \int \xi_i(\vec{r}_1) \Psi^*(\vec{r}_1, \vec{r}_2 \ldots \vec{r}_A) \Psi(\vec{r}_1', \vec{r}_2 \ldots \vec{r}_A)$$

$$\xi_j(\vec{r}_1') d^3 r_1 d^3 r_1' d^3 r_2 \ldots d^3 r_A, \quad (3.1)$$

and diagonalize it to obtain the natural orbits ψ_i and occupation numbers n(i) (Eq. (1.15)). Preliminary results of calculations of ψ_i and n(i) of a seventy-particle Bose liquid ^{4}He drop are available, and we will discuss them. Calculations of ψ_i and n(i) in seventy-particle Fermi liquid ^{3}He drop are in progress.

The natural orbits of a spherically symmetric system can be labeled with quantum numbers n, ℓ and m. Some of the natural orbits of A = 70 Bose liquid ^{4}He drop are shown in Fig. 12 along with the mean field orbit $\phi_{0,0,0}$. In mean field theory the ground state of a Bose system is obtained by putting all the A particles in the lowest energy $\phi_{0,0,0}$ state and Eq. (1.16) then gives:

$$\phi_{0,0,0}(r) = \sqrt{\rho(r)/A}. \quad (3.2)$$

The occupation numbers of some of the orbits are given in Table I. In Bose systems the fraction of particles in the $\psi_{0,0,0}$ state is called the condensate fraction n_c,

$$n_c(A) = n(0,0,0)/A. \quad (3.3)$$

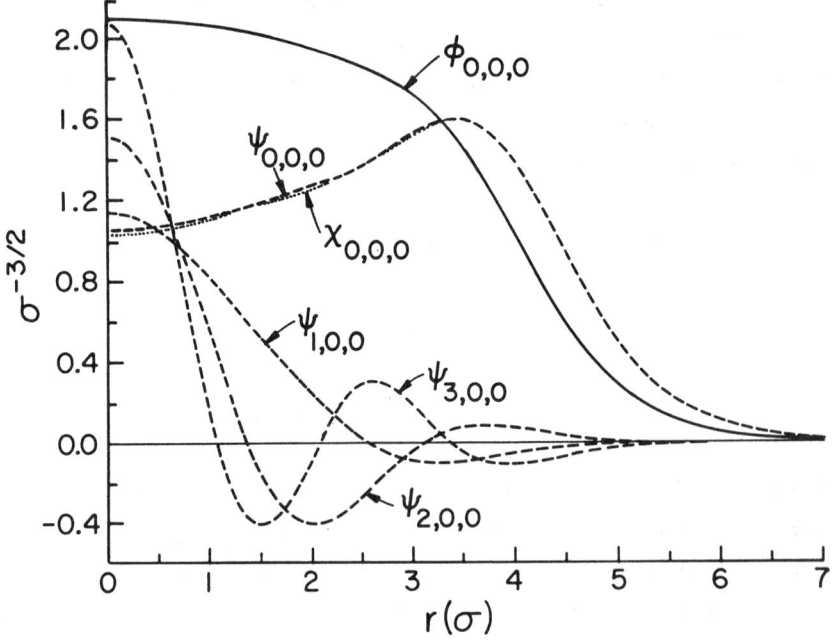

Fig. 12 The $\ell = 0$ natural orbits $\psi_{n,0,0}$ in a drop of Bose liquid ^{4}He containing seventy atoms. The $\phi_{0,0,0}$ and $\chi_{0,0,0}$ are mean-field and quasi-particle orbits.

Table I Occupation numbers of natural orbits in A = 70 Bose liquid ^{4}He drop

Orbit (n,ℓ)	Occupation number
0,0	25.3
0,1	0.49
1,0	0.44
0,2	0.43
0,3	0,37
1,1	0.35

Its value in the A = 70 drop is 0.36; it is much larger than the $n_c \sim 0.1$ estimated for extended liquid.[31] The $n(n,\ell,m \neq 0,0,0)$ are quite small, however these states contain the majority of the particles, and account for most of the density in the center of the drop (Fig. 13).

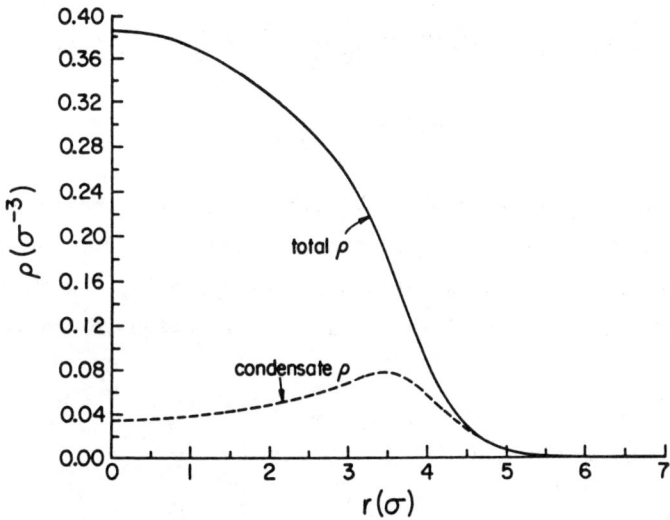

Fig. 13 The total $\rho(r)$ and the partial density due to particles condensed in the $\psi_{o,o,o}$ state.

We find that, to a very good approximation, the natural orbit $\psi_{o,o,o}(r)$ is given by:

$$\psi_{o,o,o}(r) \propto (1-0.69\ \rho(r)/\rho_o)\ \sqrt{\rho(r)}$$

$$\propto \sqrt{n_c(\rho(r))\rho(r)}. \qquad (3.4)$$

This relation can be understood from the local density approximation by identifying $(1-0.69\ \rho/\rho_o)^2$ as the condensate fraction $n_c(\rho)$ in uniform liquid at density ρ. We have in the local density approximation:

$$\rho_1(\vec{r}_1 + \frac{\vec{r}}{2},\ \vec{r}_1 - \frac{\vec{r}}{2}) \approx n_c(\rho(\vec{r}_1))\rho(\vec{r}_1) \text{ for } |r| \sim \ell, \qquad (3.5)$$

where $\ell > 1\sigma$ (the $\rho_1(r_1,r_1')$ in liquid ^{4}He approaches its asymptotic value determined by $n_c(\rho)$ when $|r_1-r_1'| > 1\sigma$ (Ref. 31)), and ℓ is less than the length scale over which $\rho(r)$ varies. The density matrix is also approximately given by:

$$\rho_1(\vec{r}_1 + \frac{\vec{r}}{2}, \vec{r}_1 - \frac{\vec{r}}{2}) \approx n(0,0,0) \, \psi_{0,0,0}(\vec{r}_1 + \frac{\vec{r}}{2}) \, \psi_{0,0,0}(\vec{r}_1 - \frac{\vec{r}}{2})$$

$$\approx n(0,0,0) \, \psi^2_{0,0,0}(\vec{r}_1), \quad |r| \sim \ell, \quad (3.6)$$

and Eq. (3.4) follows from Eqs. (3.5) and (3.6). We note that at $\rho > \rho_o$, where extended liquid exists under positive pressure, the $n_c(\rho)$ is well approximated by $(1-0.69 \, \rho/\rho_o)^2$, and also the calculated value 0.36 of the condensate fraction in A = 70 drop is close to:

$$\frac{1}{A} \int (1-0.69 \, \rho(r)/\rho_o)^2 \, \rho(r) \, d^3r = 0.37. \quad (3.7)$$

Thus it appears reasonable to identify $(1-0.69 \, \rho/\rho_o)^2$ with $n_c(\rho)$.

The quasi-particle orbit $\chi_{0,0,0}(r)$ is defined as:

$$\chi_{0,0,0}(\vec{r}) \propto \langle \Psi_{A-1} | a(\vec{r}) | \Psi_A \rangle, \quad (3.8)$$

where $|A-1\rangle$ is the ground-state of the system with one less particle, and $a(\vec{r})$ is an annihilation operator. It has also been calculated[29] for A = 70, and it is practically indistinguishable from the $\psi_{0,0,0}$ as shown in Fig. 12.

There is no known simple method to generate the natural orbits having n, ℓ, m $\neq$ 0,0,0 in Bose liquid drops. It should be pointed out here that all natural orbits are localized

$$\psi_{n,\ell,m}(r \to \infty) = 0 \text{ for all } n,\ell,m, \quad (3.9)$$

by virtue of Eq. (1.16). As a matter of fact, since the radius of the density distribution

$$\sum_{n,\ell,m > 0,0,0} n(n,\ell,m) \, \psi^2_{n,\ell,m}(r)$$

is smaller than that of $\rho(r)$ [Fig. 13], the radii of $\psi_{n,\ell,m \neq 0,0,0}^2$ are smaller than that of $\psi_{o,o,o}^2$.

The momentum distribution of particles in the liquid drop is given by:

$$n(\vec{k}) = \sum_{n,\ell,m} n(n,\ell,m) |\tilde{\psi}_{n,\ell,m}(\vec{k})|^2. \qquad (3.10)$$

It is generally calculated directly from the $\rho_1(\vec{r},\vec{r}')$, and is shown in Fig. 14 for the A = 70 Bose liquid ^{4}He drop. The n(k) has structure at small k riding on a smooth background. This structure is due to the preferentially occupied $\psi_{o,o,o}$ orbit. The large peak at k = 0 will become $n_c(\rho_o)A\delta(\vec{k})$ in the limit $A \to \infty$. The n(k) can be accurately explained as:

$$n(k) = n(0,0,0) \, |\tilde{\psi}_{o,o,o}(k)|^2 + n_{bg}(k), \qquad (3.11)$$

where $n_{bg}(k)$ is the momentum distribution due to all the $n,\ell,m \neq 0,0,0$ states (Fig. 14).

Let us now assume that the natural orbit $\psi_{o,o,o}$ is not known, as is generally the case, and approximate the n(k) by:

$$n(k) = n'(0,0,0) \, |\tilde{\phi}_{o,o,o}(k)|^2 + n_{bg}(k), \qquad (3.12)$$

where $\tilde{\phi}_{o,o,o}$ is the mean-field orbit $\propto \sqrt{\rho(r)}$. This approximation does not do as well as Eq. (3.11) becuase the width of the central maximum of $|\tilde{\phi}_{o,o,o}|^2$ is too large by 10%, its second maximum is too small by a factor of ~ 5, and at a larger value of k (Fig. 14). Nevertheless a fit to n (k $\sim$ 0) with Eq. (3.12) is not too bad, and it gives n'(0,0,0) = 33.5, to be compared with the true occupation number n(0,0,0) = 25.3 of the natural orbit. Thus it appears that it is necessary to know the natural orbit $\psi_{o,o,o}$ to extract its occupation number from n(k).

There have been a couple of attempts to estimate the natural orbits in lead nuclei[32,33] from the optical potential. Their occupation numbers have been estimated[34] by using the

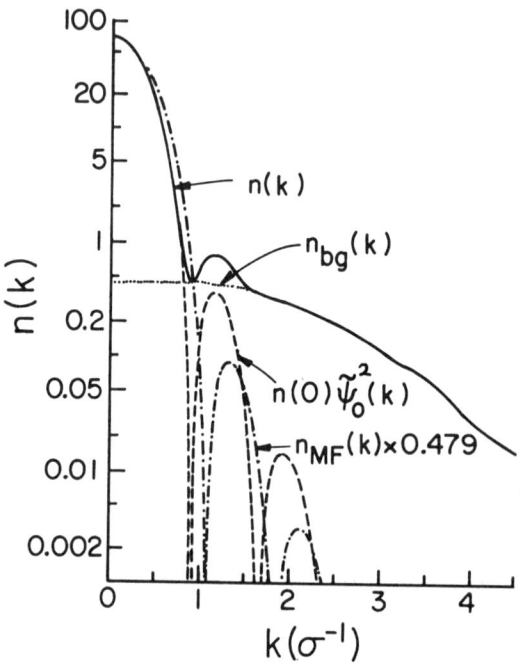

Fig. 14 The momentum distribution n(k) of atoms in an A = 70 drop of liquid ^{4}He. The dashed and dash-dot curves show n(0,0,0) $|\tilde{\psi}_{o,o,o}(k)|^2$ and n'(0,0,0) $|\tilde{\phi}_{0,0,0}(k)|^2$.

momentum distribution of nucleons in nuclear matter[35], and resuls of RPA calculations[36] to take into account surface effects. These estimates of the occupation numbers may have errors of ~ 20%, and are shown in Fig. 15. The most important quantity is the discontinuous change Z in occupation numbers at the Fermi energy, and it is estimated to be ~ 0.6 ± 0.1 in lead nuclei. The effective mass of nucleons in nuclear matter is enhanced at Fermi energy by a factor 1/Z, and the observed enhancement[37] also indicates that Z ~ 0.7 in nuclei.

The form factor of several transitions in ^{208}Pb and ^{207}Pb have been measured by e,e' reactions. Some of these transitions are expected to be dominantly single particle type. A list of the experiments and their results can be found in Ref. 34; they have been reviewed by Heisenberg[38] and Papanicolas.[39] The shape

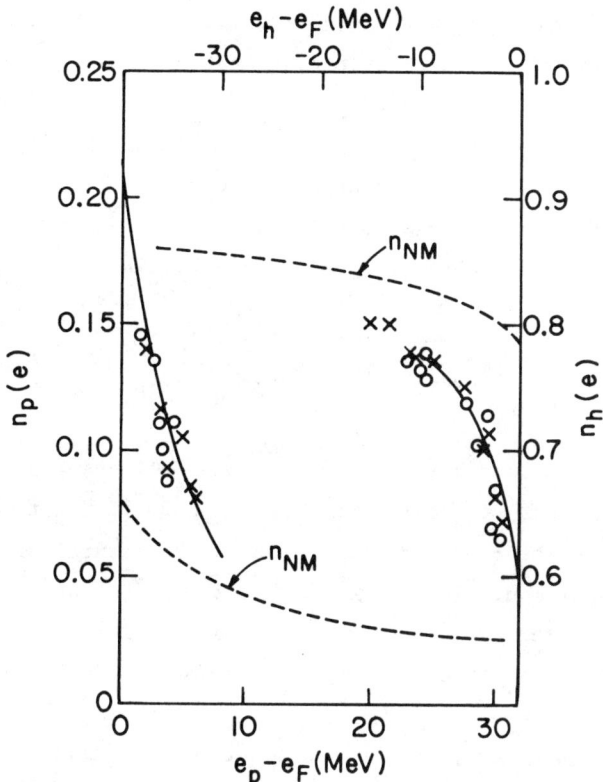

Fig. 15 Estimated occupation numbers of plane wave states in nuclear matter (dashed lines) and shell model orbits in lead nuclei (x-protons and o-neutrons).

of the form factor in the $k \sim 2$ fm^{-1} region is similar to that expected from a single-particle transition, however the magnitude is smaller than expected, i.e. it is quenched. Theoretically one expects the single-particle contributions to be quenched by a factor Z, and the observed quenching also suggestes that $Z \sim 0.65 \pm .1$ in lead nuclei.

In mean-field theory the charge difference between ^{206}Pb and ^{205}Tl is given by:

$$\rho_c(^{206}\text{Pb}) - \rho_c(^{205}\text{Tl}) = \rho_c(3S1/2) + \delta\rho, \qquad (3.13)$$

where $\rho_c(3S1/2)$ is the charge distribution of the 3S1/2 orbit in ^{206}Pb and $\delta\rho$ is the polarization contribution arising from the difference in the single particle wave functions in ^{206}Pb and ^{205}Tl. This charge difference has been measured[40], and the $\delta\rho$ has been estimated by density-dependent Hartree-Fock calculations. The single particle contribution in the observed charge difference is found to be quenched by ~ 0.7. In microscopic theories we have:

$$\rho_c(^{206}Pb) - \rho_c(^{205}Tl) = Z\rho_c(3S1/2)_{no} + \delta\rho + \Delta\rho, \qquad (3.14)$$

where $\Delta\rho$ is the contribution arising from small changes in the occupation numbers of orbits other than the 3S1/2 orbit, and $\rho_c(3S1/2)_{no}$ is the charge distribution of the natural orbit. Thus this experiment also suggests that Z = 0.7, however note that the quenching factor 0.7 is obtained in Ref. 40 from the $\rho_c(3S1/2)$ of a density-dependent Hartree-Fock orbit which may not equal the natural orbit.

^{205}Tl is not a closed shell nucleus, and hence it is possible that its ground state is a mixture of states. For the sake of simplicity we consider only the 2d3/2 orbit as the other possibility of the proton hole, however the arguments are more general. Let us assume that the ^{205}Tl ground state is given by:

$$|^{205}Tl\rangle = (1-\alpha^2)^{1/2}|(3S1/2)^{-1}\rangle + \alpha|(2d3/2)^{-1} + 2^+\rangle \qquad (3.15)$$

We then have the sum of the occupation numbers n(3S) of the two 3S1/2 orbits in ^{206}Pb and ^{205}Tl as 2 and $1 + \alpha^2$ in this mean-field theory which admits configuration mixing in the open shell. The charge difference is then given by:

$$\rho_c(^{206}Pb) - \rho_c(^{205}Tl) = (1 - \alpha^2)\, \rho_c(3S1/2)$$
$$+ \alpha^2\rho_c(2d3/2) + \delta\rho_c, \qquad (3.16)$$

and $\alpha^2 = 0.3$ would explain the observed data. This explanation implies that the ratio

$$R_5 = \frac{n(3S)[^{205}Tl]}{n(3S)[^{206}Pb]} \sim 0.65. \qquad (3.17)$$

Recently the NIKHEF group[41] has attempted to measure the ratios R_5 and R_6,

$$R_6 = \frac{n(3S)[^{206}Pb]}{n(3S)[^{208}Pb]}, \qquad (3.18)$$

by measuring the total S-wave strength in e,e'p reaction up to an excitation energy of 5.5 MeV. They assume that the strength below 5.5 MeV is proportional to n(3S). It should be pointed out here that this cutoff energy of 5.5 MeV cannot be increased arbitrarily because at higher energies one would get strength from protons in 2S1/2 and 4S1/2 orbits. They find that R_5 = 0.49 ± .02 and R_6 = 0.83 ± .05, and assuming that $n(3S)[^{206}Pb]$ - $n(3S)[^{205}Tl]$ = 0.7 obtain:

$$n(3S)[^{208}Pb] = 1.65 \pm .12 \; (\sim 1.28)$$

$$n(3S)[^{206}Pb] = 1.37 \pm .07 \; (\sim 1.28)$$

$$n(3S)[^{205}Tl] = 0.67 \pm .06 \; (\sim 0.78).$$

The values estimated in Ref. 34 are given in parenthesis. Noting that the theoretical values have ~ 20% uncertainties, and that they are based on nuclear matter results which do not contain effects of shell closures, we find a reasonable agreement.

Finally we consider the possibility of studying the quasi-particle orbits in nuclei via e,e'p reaction to the ground state of the A-1 nucleus. If we assume that distorted wave Born approximation is valid for these reactions, they measure the quasi-particle wave function defined by Eq. (3.8). In Bose ^{4}He liquid drops it appears that the quasi-particle wave function is very close to the natural orbit, and it differs from the mean-field wave function due to correlation effects. Recently the NIKHEF group[41] has studied e,e'p reaction on ^{90}Zr to the lowest $1/2^-$ and $3/2^-$ states in ^{89}Y. They find that the data are not well explained with a mean-field 2p3/2, 2p1/2 orbits and a "conventional" optical potential. However the data can be explained with an "unconventional" optical potential that has more absorption at $r \sim 0$. It may also be possible to explain the data with a "conventional" optical potential by using quasi-particle orbits that have smaller amplitude at $r \sim 0$ than the mean field orbits. A systematic study of e,e'p reaction over many nuclei may provide valuable infromation on quasi-particle orbits in nuclei.

ACKNOWLEDGEMENTS

I would like to thank M. Brussel, D. Lewart, A. Fabrocini, S. Fantoni, B. Frois, C. Papanicolas, S. Pieper, R. Schiavilla, R. Wiringa and P. de Witt Huberts for illuminating discussions and communications. This work is supported by the United States National Science Foundation via Grant PHY 84-15064.

REFERENCES

1. J. W. Negele, Rev. Mod. Phys. 54, 913 (1982).
2. P. Hohenberg and W. Kohn, Phys. Rev 136, 8864 (1964).
3. J. M. Cavedon, Thesis de Doctoral d'Etat, Paris 1980.
4. D. Gogny,, Proc. Int. Conf., Los Alamos, USA, January 1980.
5. The Three-Body Force in the Three-Nucleon System, Lect. Notes in Physics, Vol. 260, Edited by B. L. Berman and B. F. Gibson.
6. V. R. Pandharipande, Nucl. Phys. A446, 189c (1985).
7. C. R. Chen, G. L. Payne, J. L. Friar and B. F. Gibson, Phys. Rev. C33, 1740 (1986) and private communication.

8. R. Schiavilla, V. R. Pandharipande and R. B. Wiringa, Nucl. Phys. A449, 219 (1986).
9. B. D. Day and R. B. Wiringa, Phys. Rev. C32, 1057 (1985).
10. J. Carlson and M. H. Kalos, Phys. Rev. C32, 2105 (1985).
11. H. Kummel, K. H. Luhrmann and J. G. Zabolitzky, Phys. Rep. 36C, 1 (1978).
12. H. E. Conzett et al, Phys. Rev. Lett. 43, 572 (1979).
13. S. Sen and L. D. Knutson, Phys. Rev. C26, 257 (1982).
14. B. C. Karp et al, Phys. Rev. Lett. 53, 1619 (1984).
15. P. O. Löwdin, Phys. Rev. 97, 1474 (1955).
16. V. R. Pandharipande et al, Phys. Rev. Lett. 50, 1676 (1983).
17. V. R. Pandharipande, S. C. Pieper and R. B. Wiringa, Phys. Rev. B34, 4571 (1986).
18. K. E. Schmidt, Private communication.
19. T. de Forest, Jr. and J. D. Walecka, Advances in Phys. 15, 1 (1966).
20. R. Schiavilla et al, University of Illinois Preprint ILL-(NU)-86-60.
21. I. Sick, Proc. of this conference.
22. R. B. Wiringa, R. A. Smith and T. L. Ainsworth, Phys. Rev. 29C, 1207 (1984).
23. C. Marchand et al, Phys. Lett. 153B, 29 (1985), CEN Saclay report CEA-N-2439 and private communication.
24. R. Schiavilla and V. R. Pandharipande, Private communication.
25. P. Barreau et al, Nucl. Phys. A402, 515 (1983).
26. Z. E. Meziani et al, Phys. Rev. Lett. 52, 2130 (1984).
27. S. Fantoni and V. R. Pandharipande, University of Illinois Preprint ILL-(NU)-86-52.
28. C. C. Blatchley et al, Phys. Rev. C34, 1243 (1986).
29. J. Noble, Phys. Rev. Lett. 42, 412 (1981);
    C. M. Shakin, Nucl. Phys. A446, 323C (1985).
30. S. Pieper, Private communication.

31. E. Manousakis, V. R. Pandharipande and Q. N. Usmani, Phys. Rev. 31, 7022 (1985).
32. M. Jaminon, C. Mahaux and H. Ngô, Nucl. Phys. A440, 228 (1985).
33. B. L. Friman, Proc. Int. Conf. on Unified Concepts of Many-Body Problems, Stony Brook (1986).
34. V. R. Pandharipande, C. N. Papanicolas and J. Wambach, Phys. Rev. Lett. 53, 1133 (1984).
35. S. Fantoni and V. R. Pandharipande, Nucl. Phys. A427, 473 (1984).
36. D. Gogny, Lect. Notes in Physics 108, 88 (1979).
37. C. Mahaux et al, Phys. Rept. 120C, 1 (1985).
38. J. Heisenberg, Proc. of this conference.
39. C. N. Papanicolas, AIP Conf. Proc. 142, (1986), Editor H. Nann.
40. B. Frois, Nucl. Phys. A396, 409C (1983).
41. P. K. A. de Witt Huberts, Proc. Int. Nucl. Phys. Conf. Harrogate (1986).

SESSION C

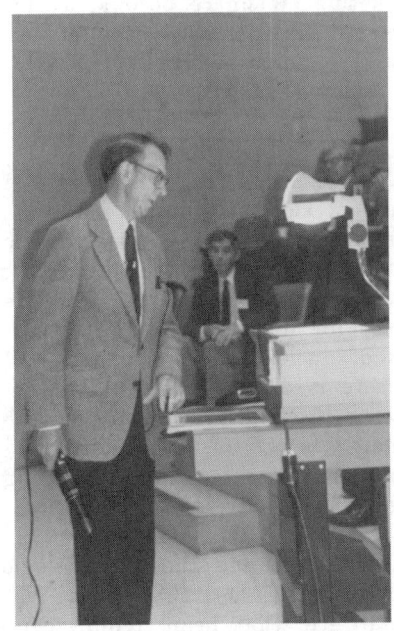

ELECTRON ACCELERATORS FOR RESEARCH AT THE FRONTIERS OF NUCLEAR PHYSICS

Hermann A. Grunder, Beverly K. Hartline, and Steven T. Corneliussen
Continuous Electron Beam Accelerator Facility (CEBAF)
Newport News, Virginia 23606

ABSTRACT

Electron accelerators for the frontiers of nuclear physics must provide high duty factor (>80%) for coincidence measurements; few-hundred-MeV through few-GeV energy for work in the nucleonic, hadronic, and confinement regimes; energy resolution of ~10^{-4}; and high current ($\geq$100 μA). To fulfill these requirements new machines and upgrades of existing ones are being planned or constructed. Representative microtron-based facilities are the upgrade of MAMI at the University of Mainz (West Germany), the proposed two-stage cascaded microtron at the University of Illinois (U.S.A.), and the three-stage Troitsk "polytron" (USSR). Representative projects to add pulse stretcher rings to existing linacs are the upgrades at MIT-Bates (U.S.A.) and at NIKHEF-K (Netherlands). Recent advances in superconducting rf technology, especially in cavity design and fabrication, have made large superconducting cw linacs become feasible. Recirculating superconducting cw linacs are under construction at the University of Darmstadt (West Germany) and at CEBAF (U.S.A.), and a proposal is being developed at Saclay (France).

INTRODUCTION

At a conference celebrating 35 years of electron physics made possible by past generations of electron accelerators, it is timely to look toward the future to identify the electron beam parameters, accelerator technologies, and facilities needed to continue to advance the frontiers of physics. Here we perform this survey for nuclear physics, where electrons have served as precise probes of nuclear and nucleon structure. The information presented here is discussed in the context of a broadly based, future nuclear physics program that includes advanced accelerators providing high-energy heavy ion and hadronic probes to obtain essential insights into nuclear matter that cannot be obtained using electron machines.[1] These other accelerators are not discussed here.

Before describing future electron accelerators and their technologies, it is appropriate to consider briefly the physics issues that these machines will be built to address, and to see what beam requirements they set. In the past, electron scattering experiments have provided precise information about nuclear wave functions, excited states, charge densities, and magnetization densities. Increasingly fine energy resolution achieved by existing electron accelerators has made it possible to isolate individual quantum states. However, coincidence measurements, in which the scattered electron and one or more nucleons or mesons it has knocked from the nucleus are studied in coincidence, provide more details about nuclear structure than can be obtained from inclusive scattering. This recent emphasis has resulted in a push for electron accelerators capable of delivering continuous beams, which greatly facilitate coincidence studies.

In parallel with the demand for more complete characterization of the scattering interaction has come an interest in even smaller subnuclear constituents. A particular focus is the transition between the nucleon-meson and the quark-gluon descriptions of nuclear matter. This interest is pushing the required electron energy higher -- to the few GeV range -- while unresolved issues related to nuclear structure on coarser scales will continue to be tackled.

Three major nuclear regimes, corresponding to different accelerator energy ranges, have been identified.[2]

1. <u>Nucleonic</u>: Spatial resolutions of a few fm, where the nucleus behaves as a collection of nucleons (electron momentum transfer up to a few hundred MeV/c).
2. <u>Hadronic</u>: Spatial resolution of the order of 1 fm, where the role of mesons and excited nucleon states is important (electron momentum transfer between a few hundred MeV/c and a few GeV/c).
3. <u>Confinement</u>: Spatial resolution less than 0.1 fm, where the details of nucleon structure, interactions between the nuclear medium and the nucleon, and quark confinement can be studied (electron momentum transfer above 1 GeV/c).

To explore these regimes requires some new and some upgraded electron accelerators to span the energy range between a few hundred MeV and several GeV. These accelerators will offer a combination of features: a high duty factor ($\gtrsim 80\%$), fine energy resolution ($\sim 10^{-4}$), ample beam current ($\gtrsim 100$ μA), and excellent beam quality.

Accelerator technology has kept pace with these requirements. There are now three viable approaches for producing high duty factor or continuous wave (cw) electron beams: microtron, linac with pulse stretcher ring, and superconducting cw linac. A microtron provides high duty factor and excellent beam quality with high energy resolution, but is limited in its maximum energy. A room-temperature (normal conducting) linac with pulse stretcher ring (PSR) offers high energy, high current, high but modulated duty factor, and modest beam quality. A superconducting cw linac can provide high energy, high duty factor, excellent energy resolution, and high current.

In this paper, which updates the reviews by Herminghaus[3] and Flanz,[4] the capabilities of operating electron accelerators for nuclear physics are summarized, generic design issues associated with each accelerator approach are discussed, and the features of representative planned electron accelerators for nuclear physics are described.

OPERATING ELECTRON ACCELERATORS

Table 1 lists the major operating electron accelerators for nuclear physics. Figure 1 graphically compares these facilities in terms of energy, duty factor, and current. Existing machines provide either high energy or high duty factor, but not both. The six machines with high energy but very low duty factor are pulsed, room-temperature linacs. Of the four low-energy, high-duty-factor machines, Mainz and Illinois are microtrons, Lund is a microtron with PSR, and Sendai is a linac with PSR.

FUTURE ELECTRON ACCELERATORS

Several laboratories and universities have developed and proposed designs for electron accelerators to access one or more of the three main nuclear regimes: the nucleonic, hadronic, and confinement regimes. These machines are listed in Table 2. Their capabilities are presented graphically in Figure 2, which illustrates the uniform quest for high duty factor throughout the energy ranges of interest.

Four of the machines are to be microtrons: Illinois, NBS (National Bureau of Standards), the Mainz upgrade of MAMI (MAinz MIcrotron), and the Troitsk-Lebedev "polytron." Illinois is awaiting a funding decision, the Mainz upgrade is in progress, and a prototype first stage for Troitsk is under construction in Moscow. The NBS microtron was terminated in late 1986 as a nuclear physics project, and is being converted to a free electron laser (FEL) facility. NIKHEF-K in Holland, MIT-Bates, and EROS (Electron Ring of Saskatoon, Canada) have proposed pulse stretcher ring (PSR) additions to existing pulsed linacs. The PSR at EROS is nearly operational, while the other projects are awaiting final funding decisions. Of the four superconducting linacs, two are under construction (the Continuous Electron Beam Accelerator Facility (CEBAF) in Virginia and the University of Darmstadt in West Germany), one is being designed to replace the pulsed Accelerateur Lineare de Saclay (ALS) linac in France, and one is under consideration at Frascati, Italy. The low-duty-factor Frascati internal target experiment and the two low-current Bonn machines shown on the figure are synchrotrons.

Representative microtron, linac-PSR, and superconducting linac projects are discussed in the following sections, after a brief discussion of the characteristics and issues associated with each design approach.

Microtrons

A microtron[5] is an energy-efficient approach to producing energetic particle beams, because the accelerating structure is reused many (N) times. In a classical microtron, the accelerating structure consists of a single rf cavity placed in a uniform magnetic field (Figure 3).

Table 1
Operating Electron Accelerators for Nuclear Physics

Machine	Location	Type	Energy (MeV)	Average Current (μA)	Duty Factor (%)
Illinois	Urbana, IL, USA	microtron	100	10	100
MAX	Lund, Sweden	microtron/PSR	100	30	≥90
Tohoku	Sendai, Japan	linac/PSR	150	3	90
*MAMI	Mainz, W. Germany	microtron	180	100	100
EROS	Saskatoon, Canada	linac	300	70	0.1
Frascati	Frascati, Italy	linac	400	80	1
MEA/NIKHEF-K	Amsterdam, Netherlands	linac	530	60	1
ALS	Saclay, France	linac	720	100	1
*Bates/MIT	Middleton, MA, USA	linac	850	100	1
*SLAC/NPAS	Palo Alto, CA, USA	linac	4000	15	0.03

*Polarized beam available

Table 2
Future Electron Accelerators for Nuclear Physics

Machine	Location	Type	Max. Energy (GeV)	Average Current (μA)	Duty Factor (%)	Status
Darmstadt	W. Germany	recirc SC linac	0.13	20	100	Under construction
Lebedev	Moscow, USSR	microtron	0.14	100	100	Under construction
NBS	Gaithersburg, MD, USA	racetrack microtron	0.20	200	100	Under construction
EROS	Saskatoon, Canada	linac/PSR	0.30	70	≥80	PSR under commissioning
Illinois	Urbana, IL, USA	cascaded microtron	0.45	100	100	Proposed
Update/NIKHEF	Amsterdam, Neth.	linac/PSR	0.7	40	≥80	Proposed
*MAMI	Mainz, W. Germany	cascaded microtron	0.84	100	100	Under construction
*Bates/MIT	Middleton, MA, USA	linac/PSR	1.0	100	≥85	PSR proposed
	Frascati, Italy	SC linac	~1.0	?	100	Under consideration
ADONE	Frascati, Italy	synchrotron	1.5	80	1-5	Internal target capability under construction (bremsstrahlung) proposed (scattering)
ALS-II	Saclay, France	recirc SC linac	1.5-2	100	100	Design in progress
*ELSA	Bonn, W. Germany	synchrotron/PSR synchrotron/booster	2.3 3.5	0.2 0.04	95 20	Construction nearly complete
*CEBAF	Newport News, VA, USA	recirc SC linac	4.0	200	100	FY 1987 start (?)
Lebedev	Troitsk, USSR	cascaded polytron	4.5	100-200	100	Proposed

*Polarized electrons available

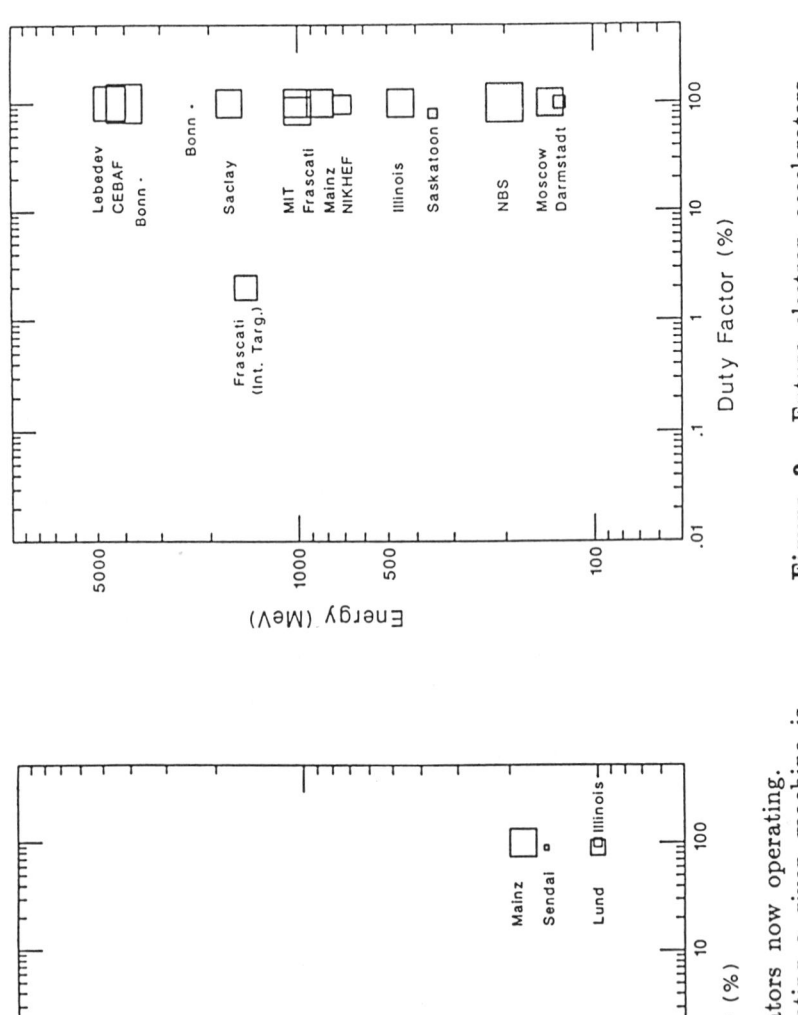

Figure 1. Electron accelerators now operating. The area of the box representing a given machine is proportional to the machine's current. (For reference, the Mainz box represents a current of 100 μA.)

Figure 2. Future electron accelerators.

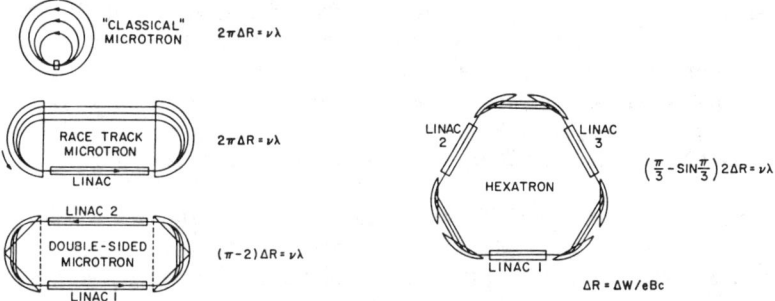

Figure 3. Variations on the microtron.[6]

On sequential recirculations the beam moves through circular orbits of increasing radius. The orbits share a common tangent at the cavity. The microtron resonance condition requires that the time difference between successive orbits be an integral multiple of the rf period to ensure that the beam reenters the cavity in phase on each pass. For fully relativistic particles, the path-length difference (ΔL) between successive orbits must be an integral multiple of the rf wavelength (λ).

$$\Delta L = n\lambda \qquad (n=1, 2, ..)$$

Usually n is less than 5 to optimize phase acceptance.

In a racetrack microtron, the magnetic field is split into two halves, which are separated and have a linac section placed between them (Figure 3). Particles trace elongated, racetrack-shaped orbits, which all pass through the linac. On successive recirculations the radius of the beam trajectory through the end magnets increases, thereby increasing the width of the racetrack orbit. The increased path length must still obey the resonance condition given above, and quite stringent requirements are placed on the field uniformity of the end magnets. The Illinois and Mainz accelerators are racetrack microtrons.

By splitting the end magnets and recombining the beams for acceleration through a second linac on the return path, a double-sided microtron can be built (Figure 3). Multisided polytrons with three or more linac segments separated by split "end" magnets have also been designed.[6,7] The Troitsk machine described below is an example.

The beam energy achieved by a microtron is $N \cdot \Delta E + E_{inj}$, where ΔE is the energy gain per pass and E_{inj} is the injection energy. Typically N is between 20 and 90. To achieve an energy higher than achievable by one microtron, present proposals call for cascaded machines, where the extracted beam from one microtron is injected into a second one (and possibly a third) for further acceleration. All three proposals discussed here are for cascaded microtrons.

Advantages of microtrons include energy efficiency, an ability to extract several beam energies simultaneously by extracting portions of the beam from different orbits, smooth (truly cw) macroscopic time structure, with microscopic time structure corresponding to the rf frequency, and excellent beam quality at low energies where quantum excitation is small. In addition, microtrons have intrinsic phase stability.

The major design issues for microtrons are accurate beam control and correction, beam breakup, and emittance growth due to synchrotron-radiation-induced quantum excitation. These issues are treated briefly below.

Accurate beam control and careful correction of errors in the beam transport system are required to maintain the appropriate phase between the rf accelerating voltage and the beam bunches and to maintain the position of the beam in the bore of the structure.

Because the total time that electrons spend in the microtron is extremely short compared with the time they would spend in a storage ring, instabilities characteristic of storage rings do not occur in microtrons. Only two types of beam instabilities tend to limit microtron currents.[8] The first, cumulative beam breakup, occurs when an electron bunch passes off center through the accelerating structure and excites transverse modes. These transverse fields deflect other electron bunches, which excite the modes further. The second instability, regenerative multipass breakup, occurs when the transverse modes excited by one electron bunch deflect the same bunch on its subsequent passes through the linac structure. These instabilities are not a problem, at present intensity levels, in microtrons with room-temperature accelerating structures, but appear to have limited the current of microtrons with superconducting accelerating structures.[9] Apparently the low Q's of the copper structures effectively damp the disruptive modes.

Above a few GeV, synchrotron radiation tends to increase the emittance of electron beams significantly. The emission of this radiation occurs in randomized energy quanta, so there is a spread in the radiated energy. This energy spread translates into growth in both the transverse and longitudinal emittances. At high energies, this quantum excitation determines the beam emittance. Thus, it is a major factor in choosing the accelerating-structure aperture, and may limit the maximum energy achievable with microtrons to a few GeV.

The Mainz microtron upgrade, the Illinois proposal, and the Troitsk polytron are described below. All are cascaded microtrons.

Mainz (MAMI). The final stage of a three-stage racetrack microtron is under construction at the Institute for Nuclear Physics of the University of Mainz, West Germany (Figure 4). When completed, MAMI-B will produce an 840-MeV, 100%-duty-factor, 100-μA beam with 10^{-4} energy resolution.[10,11] Its experimental program will focus on studies of hadronic degrees of freedom.

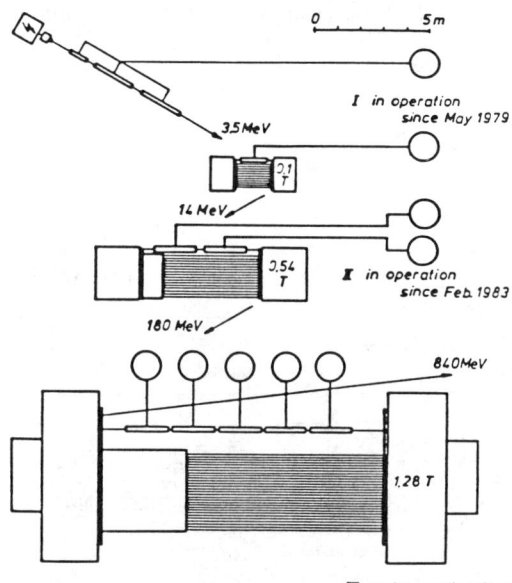

Figure 4. MAMI.

In the completed facility, electrons from a 3.5-MeV linac will be injected into Stage 1. After 18 passes through one 0.6-MeV, 80-cm section of copper accelerating structure for an output energy of 14 MeV, the beam will be sent to Stage 2, where 51 passes through a pair of 1.78-m accelerating sections providing 3.25 MeV per pass will lead to an output energy of 180 MeV. In 88 traversals of the five-section, 7.5-MeV linac of Stage 3, the beam will attain 840 MeV. The MAMI-B project includes, in addition to the third-stage 840-MeV racetrack microtron, a new injector linac to replace the present Van de Graaff. A polarized source is planned, along with a beam splitter to produce three beams simultaneously. A new experimental hall to house three spectrometers may be added later.

The MAMI facility was originally conceived in 1974. The first stage became operational in 1979, and was the first room-temperature racetrack cw microtron. The second stage was commissioned in 1983. By January 1987, the accelerator hall for Stage 3 was completed, and the first end magnet had been received. The first beam from Stage 3 is scheduled for early 1989.

<u>Illinois</u>. The Nuclear Physics Laboratory of the University of Illinois has proposed construction of a 100%-duty-factor, 100-μA two-stage cascaded microtron (Figure 5) for experiments in the energy range from 80 to 450 MeV.[9] A beam energy spread of 10^{-4} is expected. This

machine is designed to perform high-resolution studies of nucleonic and mesonic degrees of freedom.

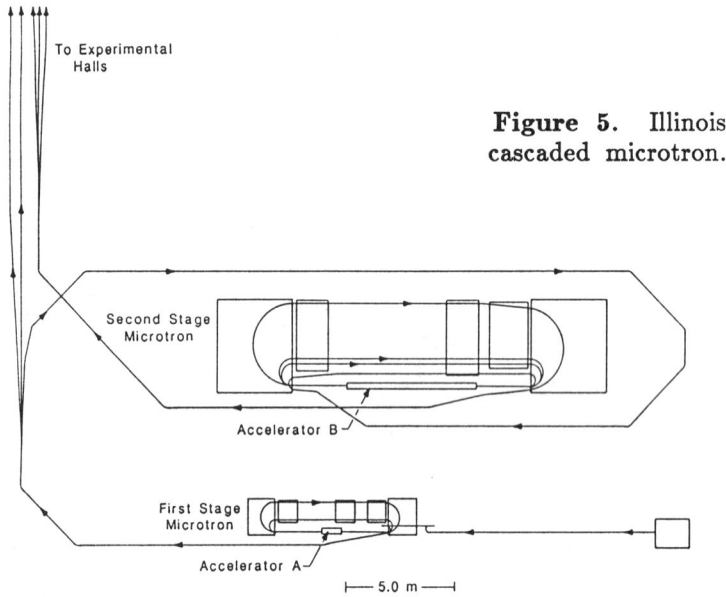

Figure 5. Illinois cascaded microtron.

Electrons injected into the first stage at 4.5 MeV will traverse a 1-m, 1.45-MeV linac 29 times for an output energy of 46.7 MeV, and then attain a final energy between 82 and 456 MeV (variable in 11.7-MeV steps) in 6 to 70 passes through the 6-m, 5.84-MeV linac of the second stage. Up to three simultaneous beams at the final energy or at a combination of the final energy and the first-stage output energy can be delivered to experimenters through the use of a subharmonic chopper in the injector and rf separators between Stages 1 and 2 and at the second-stage output. For full-beam-power experiments, the 100-μA beam is available by switching off the separator. Capability for parasitic tagged photon beams is under consideration. A capability for polarized beams is not included, but could be added, and design studies are underway.

Both the proposal and the ensuing research and development have been undertaken in the context of Illinois' experience building and operating microtrons. In 1972, MUSL-1 (Microtron Using a Superconducting Linac) at Illinois became the first operating cw racetrack microtron. A second racetrack microtron, MUSL-2, now provides 10-μA cw beams at energies up to 100 MeV. The proposal for a cascaded microtron has been favorably reviewed. Working prototypes of key components are in hand, funding for construction has been requested from the U.S. National Science Foundation, and a decision is pending for a construction start in FY 1988.

Troitsk "Polytron". Whereas the microtrons discussed above will operate below 1 GeV, where emittance growth due to synchrotron radiation is not severe, accelerator designers in the Soviet Union have proposed a 4.5-GeV cascaded "polytron" (Figure 6) for the Lebedev Physics Institute at Troitsk.[7] The physics goals include exploring the transition region and understanding quark confinement. The design calls for 100% duty factor, 300-μA current, and an energy resolution of 10^{-4}. The design has three stages: 7-MeV injection for 200-MeV output after 16 passes through the first-stage racetrack microtron, attainment of 1 GeV in 16 passes through the second-stage "quadrutron," and final energy of 4.5 GeV after 10 passes through the third-stage "octutron." The racetrack microtron includes one 8.5-m linac providing 12 MeV of acceleration. The quadrutron has two parallel 20-m linacs of 25 MeV each, and the octutron has four 77-m-long linacs providing 90 MeV each. The machine was proposed in 1984. A prototype first-stage microtron--140 MeV, 100 μA--is under construction in Moscow.

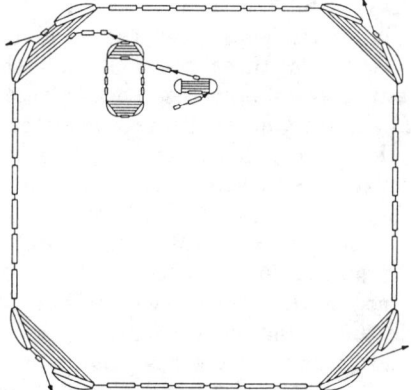

Figure 6. Troitsk "polytron."

Linacs with Pulse Stretcher Rings

Room-temperature linacs are capable of producing cw electron beams if they are operated at a low (~1 MV/m) accelerating gradient. However, both the capital cost of the required length of structure and the electric power consumption to achieve high electron energies are prohibitive. Therefore, room-temperature linacs are operated in a pulsed mode, with heavy beam loading during the pulse. Typical pulse durations are of the order of 1 to 50 μs, and the pulse rates may be of the order of 100 to 1000 Hz. The duty factors thus range from ~0.1% to a few percent. To reduce capital and operating costs, one recirculation through the linac can be employed.

By injecting the linac pulse into a ring, and extracting the beam slowly and uniformly during the interval between pulses, the duty factor from a pulsed linac can be increased to above 80%. Such a ring is known as a pulse stretcher ring (PSR) (Figure 7).[12]

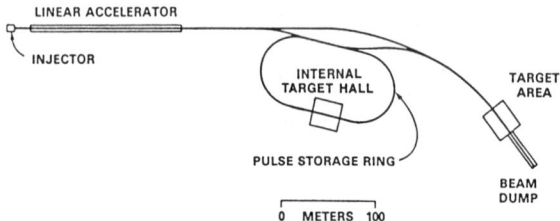

Figure 7. Generic linac-PSR.

Beam injection into the PSR is typically accomplished by single-turn, two-turn, or three-turn injection, such that the head of the electron pulse travels exactly once, twice, or three times around the ring before the tail enters. Resonant extraction is used to extract the beam during the few hundred to few thousand orbits traversed before the ring is empty and readied for the next pulse. This cycle is a rather fast "slow resonant extraction" mode, when compared with the slow extraction over a million turns from a proton synchrotron.

Pulse stretcher rings provide a cost-effective approach to producing high-duty-factor beams from existing pulsed linacs. In addition, they provide the capability for internal target experiments within the PSR. Beam quality, in terms of both emittance and energy spread, is degraded compared with that of an intrinsically cw machine, due to the high peak current in the linac. Continuous beams at only one energy can be delivered at any time, and to change energies requires resetting and retuning the PSR.

The major design issues for linac/PSR facilities are instabilities in the linac and PSR, and smooth extraction from the PSR. In addition, rf power requirements for the linac may place difficult demands on the rf system and components.

Instabilities in both the linac and PSR must be controlled. In the linac, peak currents are sufficiently high that cumulative beam breakup is the primary concern; multipass beam breakup also must be considered if recirculation is used. In the PSR, the beam must avoid the numerous instabilities that plague high-current electron storage rings.[13] However, the residence time is sufficiently short that many of these instabilities lack the time to develop and are not a problem. Unfortunately, for the same reason, essentially no beam conditioning will occur, because the time constant for beam damping significantly exceeds the residence time.

Of major importance is precise control and careful design of the extraction process. To ensure that the extracted beam is uniform over time, careful attention must be paid to the resonant extraction system and its feedback control. Modulations in the intensity of the extracted beam reduce its duty factor.

The proposals by MIT-Bates and by NIKHEF-K in Holland to add PSR's to their operating pulsed linacs are described below. PSR construction at EROS in Saskatoon, Canada has been completed; commissioning is expected to continue through mid-1987, at which time its 100- to 300-MeV, 70-μA beam at 80% duty factor will be available for experiments.[14]

MIT-Bates Upgrade. Presently operating at MIT-Bates is an 850-MeV, 1%-duty-factor, recirculating, pulsed linac. The key elements in the proposed upgrade of this machine are a PSR, an internal target hall, and a recirculator extension.[15] Main beam parameters of the proposed design are 300-1000 MeV energy, ~85% duty factor, 100-μA current, and energy resolution of 4×10^{-4}. A polarized injector is being commissioned.

Figure 8 shows the general layout of the upgrade. The PSR operates on one- or two-turn injection. The internal target hall intersecting the PSR allows use of the 40-80 mA circulating current for experiments with gaseous or very thin targets. An energy compression system has been incorporated to reduce the energy spread prior to injection into the PSR. The recirculator extension improves the machine's performance at high energy by allowing a head-to-tail recirculation of a beam pulse suitable for two-turn injection. Besides providing for 100-μA beams at energies above 500 MeV, the extension improves beam quality and operational reliability.

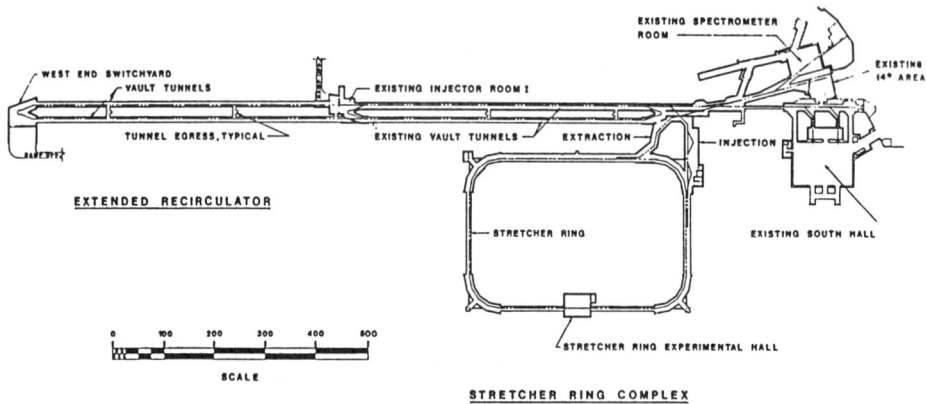

Figure 8. MIT-Bates linac-PSR.

The machine, as upgraded, would continue MIT's tradition of leadership on high-resolution studies of hadronic degrees of freedom in nuclei. It has been proposed for funding to the U.S. Department of Energy and endorsed by the U.S. Nuclear Science Advisory Committee. Preconstruction R&D is underway, and a decision on construction is pending.

NIKHEF-K. An upgrade, called Update, has been proposed for the presently operating pulsed 500-MeV, 1%-duty-factor electron linac at NIKHEF-K in Amsterdam[16] (Figure 9). Plans are for a PSR to be added, and for its injection energy--the output energy of the present linac--to be raised to 700 MeV by upgrading four or five of the linac's twelve klystrons. The PSR will use three-turn injection. To fill the PSR, the linac will operate with a shorter pulse length than at present, and at a duty factor of 0.1%. Since the linac will accelerate a higher peak current than at present, an energy compressor will be installed between the linac and the PSR to reduce the energy spread for injection into the PSR. The new performance parameters will be 15-700 MeV energy, 40 μA average current, >80% duty factor, and 5×10^{-4} energy resolution. During the fall of 1986, the proposal was reviewed favorably, and a funding decision is pending by the Dutch government.

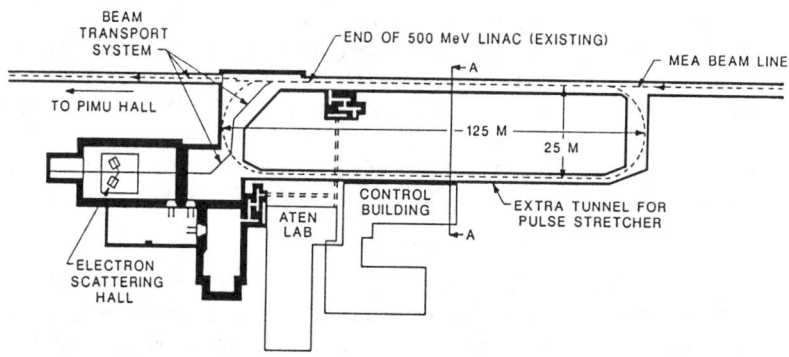

Figure 9. NIKHEF-K linac-PSR.

Superconducting CW Linacs

Superconducting linac structures have a significantly higher Q and therefore a significantly higher shunt impedance than room-temperature structures. This gain of a factor of 10^5 translates into a comparable decrease in power loss in the structure. Over 99% of the rf power goes into the beam, and only a small amount heats the walls.

To achieve this high rf efficiency, the accelerating structure must be kept superconducting. A liquid helium refrigerator is necessary to maintain these temperatures and to remove the rf-generated heat. Such a refrigerator operates at an efficiency of around 0.1%. Thus to remove 1 kW of heat requires around 1 MW of AC power for the refrigerator. Whereas it takes a few hundred MW of rf power to produce a 1-MW beam from a room-temperature cw linac, it takes only a few MW, mostly for cryogenics, to drive a comparable superconducting cw linac. Temperature optimization involves trading off cavity performance and heat loss (which are better at lower temperatures), with the complexity and cost of the cryogenic system (which are better at higher temperatures). Operating temperatures for existing and proposed superconducting accelerators are in the range 1.8 K to 4.5 K, depending on the rf frequency and the cavity material.

The most straightforward cw accelerator is a single linac, which the beam traverses only once. With an accelerating gradient of 5 MV/m, the capital cost of the length of a structure required to achieve high energies makes it economically unattractive. A more cost-effective solution is obtained by passing the beam a few times through a shorter linac structure. This concept is similar to that used for the microtron; however

1. fewer recirculations are used,
2. the recirculation paths are all separate and employ strong-focusing lattices to control beam quality,
3. phase stability is not intrinsic, and

4. the energy gain per pass can be arbitrarily large.

By appropriate selection of the field strengths of the bend magnets in each recirculation path, the lengths of all paths can be made essentially equal, and the recirculation beam lines can be stacked vertically in one tunnel.

In a recirculating linac the path of a single electron bunch from injector to experimental area comprises several (N) acceleration cycles, each of which is as follows:
1. injection into the linac,
2. acceleration by the linac,
3. "spreading" to the proper beam line for recirculation, and
4. transport through the recirculator to a recombiner for reinjection into the linac.

At any time, there are electrons at N energies passing simultaneously through the linac. Since all the particles are fully relativistic there is negligible phase slip during a pass. On the final cycle, this sequence is interrupted for extraction at step 3 where the beam, after spreading, enters an extraction beam line. Extraction elements can be placed in each of the recirculation arc beam lines to allow extraction on any of the preceding cycles as well. Such an arrangement makes possible the simultaneous delivery of beams at different but correlated energies.

The key issues for a recirculating, superconducting cw linac are beam stability, beam quality, and cavity design and performance.

In a recirculating linac, as in a microtron, the beam current is limited by multipass regenerative beam breakup.[17][18] Since the Q of superconducting cavities is so high, this problem is potentially more serious than it is for room-temperature linacs. Its solution lies in designing the accelerating cavities to damp the offending transverse modes to acceptable levels.

The second issue is the problem of conserving the emittance and momentum spread. In a recirculating linac two factors can degrade the beam: synchrotron radiation during bending in the recirculation arcs, and possible phase mismatch of the electron beams upon reinjection into the linac segments. Beam quality can be maintained and reinjection mismatches avoided through proper design of the lattices in the recirculation arc beam lines. Suitable lattices are similar to those employed in low-emittance storage rings.[19] These lattices control beam path length and provide isochronicity, achromaticity, and careful correction of chromatic effects to facilitate reinjection after each recirculation. Strong focusing minimizes emittance growth caused by quantum excitation due to synchrotron radiation.

Superconducting cavity design and performance have been the subject of considerable R&D since the 1960s when a superconducting cw electron linac was first proposed.[20][21] Major R&D efforts have been underway at Stanford, Karlsruhe, Cornell, CERN, DESY, Wuppertal, Orsay, and KEK. The goal has been to achieve high Q's and high gradients, which requires controlling surface defects and cleanliness, multipacting, and field emission. In addition, the transverse modes must be suppressed to achieve beam stability. Until recently, achieved gradients were limited to about 3 MV/m. Now gradients of 5 to 7 MV/m with Q's in excess of 10^9 are achieved routinely, at laboratories and by industry (Table 3).[22] These gradients already significantly exceed the cw gradients (~1 MV/m) feasible with copper cavities. Recent progress has been very rewarding, and there are firm plans now to employ these structures in several planned accelerators and major upgrades for nuclear and high energy physics.[23]

Table 3
Performance of $\beta=1$ Superconducting RF Cavities

Laboratory	CERN			KEK	DESY	Cornell	CEBAF**	Darmstadt/Wuppertal	
Accelerator	LEP II			TRISTAN	HERA	CESR	CEBAF	130 MeV	Recyclotron
Material	Nb	Nb	Nb/Cu	Nb	Nb	Nb	Nb	Nb	Nb_3Sn
Frequency (MHz)	350	500	500	500	1000	1500	1500	3000	3000
Operating Temperature (K)	4.2	4.2	4.2	4.2	4.2	1.8	2.0	1.8	4.2
Best Single-Cell Results									
E_A (MV/m)	10.8	13.0*	10.8	7.6*	5.5	22.8*	-	23.1*	7.2
Q at E_A (× 10^9)	1.8	0.7	0.4	0.6	0.5	2.5	-	1.2	1.1
Best Multicell Results									
No. of Cells	4	5	4	3	9	5	5	5/20	5
E_A (MV/m)	7.5*	5.0	5	5.8	5.5	15.3*	12.0*	12.3/7.4	4
Q at E_A (× 10^9)	2.2	0.7	0.8	0.6	0.5	2.2	2.4	3.5/1.2	4.5

* Cavities fabricated from high-thermal-conductivity niobium
** Cornell cavity design
Source: H. Piel, Wuppertal

Since superconducting linacs are likely to become the approach of choice for producing cw electron beams for nuclear physics, it is appropriate to discuss the status of superconducting rf technology.

Cavity shape is an important factor for both cost and performance. Within the past few years, several improvements in this area have been developed. Spherical or elliptical cell shape has been shown to reduce multipacting. Couplers for fundamental power and for higher order mode suppression attach to the beam pipe to minimize field enhancement and multipacting. The number of individual resonant rf cells (half wavelength) in a cavity is limited to control HOMs.

Optimized designs call for five cells or fewer. Cavity design and HOM suppression are aided now by the availability of computer codes such as URMEL.[24]

The most common superconductor currently in use is niobium. Since only a very thin surface layer on the inside of the cavity is active in the formation of the accelerating field, the quality and cleanliness of that surface layer is of the utmost importance to cavity performance. In addition, the surface layer must be kept below the superconducting transition temperature; cooling must be adequate to remove the heat generated by rf dissipation in dust and defects.

Recent developments in niobium processing and cavity fabrication have resulted in real progress in these areas. Niobium suppliers have developed the capability to produce niobium sheet with high thermal conductivity. High thermal conductivity helps to stabilize the cavity against being driven normal by resistive heating at a defect. Titanium treatment or yttrification can be used to increase the thermal conductivity.[25] The use of clean rooms and clean manufacturing protocols prevents the introduction of dust or dirt on the active surface. Refined electron beam welding methods help achieve weld smoothness.

Another recent development is thermometric mapping[26] to locate hot spots caused by defects or dirt on the superconducting surface. A cavity can be tested and the factor limiting its gradient can be located and repaired.

Although gradients of 5 to 7 MV/m and Q's in excess of 10^9 are achieved routinely by industry today in prototype cavities, the real attraction of rf structures is their potential to achieve gradients above 20 MV/m with very low rf losses (Q $>\gtrsim 10^{10}$) and high beam-current capacity. Single-cell cavities fabricated at Cornell and Wuppertal have achieved such high gradients (Table 3), which are far below the theoretical limit. The gradient limit is determined by the magnetic field at which the superconductor goes normal, and is in the range of 50 to 100 MV/m for the superconducting materials of interest.

Experimentation is underway with Nb_3Sn, niobium on copper, and other superconductors as cavity materials. Nb_3Sn offers the potential for lower rf losses and operation at higher temperatures. Niobium on copper would have a very high bulk thermal conductivity, thus providing excellent thermal stabilization of submicroscopic defects and dust.

In summary, the capabilities of superconducting accelerator cavities have improved significantly over the past decade. Reasonable design parameters achievable with niobium cavities fabricated by industry today are a frequency between 350 MHz and 3000 MHz, an accelerating gradient of 5 to 7 MV/m, and a Q of 2×10^9 to 3×10^9, with disruptive higher order mode Q's damped to 10^4 or 10^5. For nuclear physics applications, frequencies in the upper end of the frequency range may be preferred, because the individual micropulses cannot be resolved by the detectors, even after the frequency is reduced by splitting the beam to two or three experimental areas.

Currently under construction are two recirculating, superconducting cw linacs: one at the University of Darmstadt (West Germany), and one at CEBAF in Newport News, Virginia. Saclay (France) is developing a design and proposal, and the Italian Nuclear Physics Laboratory in Frascati is considering doing so. The following descriptions summarize the Darmstadt, CEBAF, and Saclay designs.

Darmstadt. Under construction at the Institut fur Kernphysik, Technische Hochschule, Darmstadt, West Germany, is a superconducting recirculating accelerator (Figure 10) based on 1-meter-long, 20-cell accelerating cavities operating at 3-GHz frequency.[27] The design gradient is 5 MV/m and the design Q is 3×10^9 at the operating temperature of 2.0 K.

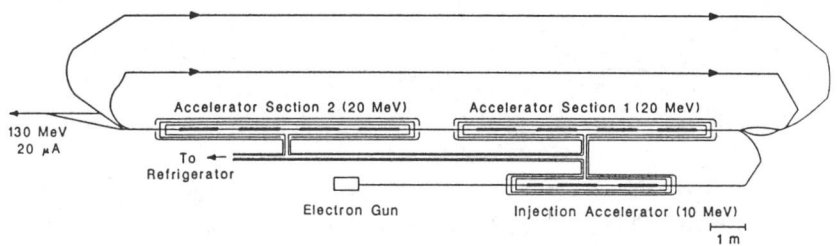

Figure 10. Darmstadt recirculating superconducting linac.

The Darmstadt electron beam is produced by a 350-keV gun and injected into one short (5-cell) and two 20-cell superconducting cavities to reach its 10 MeV injection energy. Subsequently the beam attains 130 MeV in three passes through two 20-MeV accelerating sections of four cavities apiece. The R&D work leading to the cavity design has been done in collaboration with the University of Wuppertal, which is also involved in the construction effort. The electron gun and cryogenic system are operating, and a 1.5-MeV beam has been produced using the first (short) injector cavity. Physics experiments are to begin shortly using the low-energy beam, and will continue as the machine is completed and commissioned.[28]

CEBAF. The Continuous Electron Beam Accelerator Facility (CEBAF)[29] in Newport News, Virginia, is planned as a state-of-the-art cw electron accelerator, producing a 4-GeV, 100%-duty-factor, 200-μA beam of 2×10^{-4} energy resolution. CEBAF's scientific objective is to study the structure of the nuclear many-body system, its quark substructure, and the strong and electroweak interactions prominent in the nucleus. The beam energy was specified by the U.S. nuclear physics community to access the confinement regime.

Acceleration is to take place in two antiparallel linac segments (Figure 11), with recirculation beam lines for four orbits. CEBAF's niobium accelerating cavities were developed by Cornell University's Newman Laboratory (Figure 12). They operate at 1.5 GHz and have five cells. The design specifications are a Q_o of 2.4×10^9 at a gradient of 5 MV/m and a temperature of 2.0 K. In total, the two linac segments contain 400 cavities providing 1 GeV of acceleration.

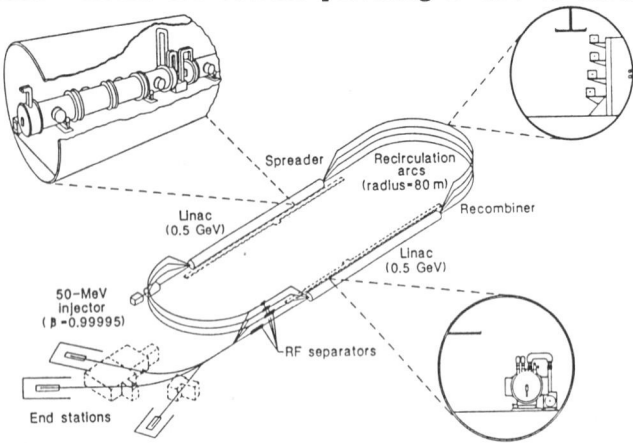

Figure 11. CEBAF recirculating superconducting linac.

Figure 12. CEBAF-Cornell five-cell niobium cavity (length = 66 cm).

In addition, there are 18 superconducting cavities in the injector, which produces a nominal 50-MeV beam. During 1986, CEBAF and Cornell worked with industry to qualify vendors to produce these cavities. To date, six prototypes, all exceeding the specifications, have been delivered and tested.

A central helium liquefier supplies liquid helium at 2.0 K to insulated cryomodules, each containing eight cavities. The extraction system directs three simultaneous beams at optionally different energies to three experimental areas.

CEBAF has been approved and funded through the U.S. Department of Energy. Construction is scheduled to begin in January 1987 and to be completed in 1992.

Saclay. In the summer of 1986 plans were advanced to replace the existing, pulsed room-temperature linac--Accelerateur Lineare de Saclay (ALS)--at Saclay, France, with a recirculating, superconducting linac.[30] Figure 13 shows the stages in the proposed evolution of ALS Supra. By 1992 a 200-meter (70-meter active length) superconducting linac, built parallel to the existing 1%-duty-factor device, would produce a 500-700 MeV beam at 100% duty factor and 100 μA. Recirculation arcs would then be installed, the ALS tunnel would be used for antiparallel transport of the beam (without further acceleration), and by 1993, three-pass, 1.5-2 GeV operation of the single superconducting linac would be possible. In later upgrades, Saclay would install a second SC linac, a new injector, additional recirculation beam lines for a final energy as high as 6 GeV, and possibly some new experimental areas.

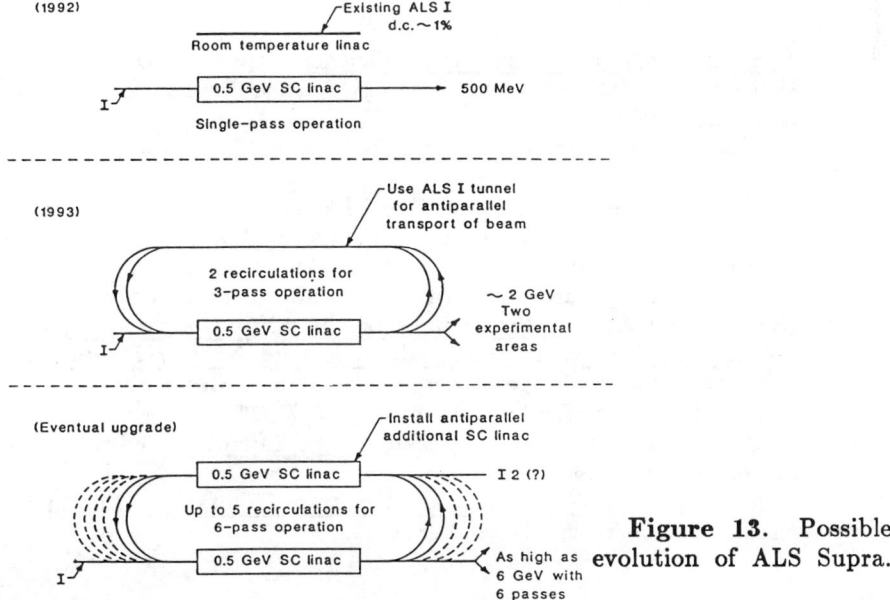

Figure 13. Possible evolution of ALS Supra.

ALS Supra is now being designed, and a cavity development program is underway, involving formal collaboration with CERN. Saclay proposes to develop four- or five-cell 1500-MHz cavities with coaxial couplers for fundamental power and for damping higher order modes. The design gradient is 7 MV/m.

In June 1986, the French nuclear physics community endorsed ALS Supra as their highest priority project.[31]

CONCLUSIONS

While the decisions on funding and construction have not been made for many of these new electron accelerators, the numerous proposals certainly presage exciting new opportunities for electromagnetic nuclear physics. It seems clear that the capabilities available globally to the experimentalist are shifting toward continuous beams, and to higher energies, while maintaining excellent energy resolution.

With the advent of superconducting rf technology as a practical means of providing high-quality cw beams, and its first major application in accelerators for nuclear physics, our field is assuming a leadership role for developing and testing accelerator technologies that have broad application. In this development, nuclear physics has the attention of the high-energy-physics and free-electron-laser communities, as well as industry. It is apparent that superconducting rf technology, conceived in the 1960s, has begun in the 1980s to deliver on its promises, and has considerable room to increase in performance and decrease in cost in the years ahead.

REFERENCES

1. A Long Range Plan for Nuclear Science, report of U.S. Department of Energy and National Science Foundation, December 1983.
2. "Report of the NSAC Subcommittee on Facility Construction," August 1986.
3. H. Herminghaus, "Survey on C.W. Electron Accelerators," in Proceedings of the 1984 Linear Accelerator Conference, edited by N. Angert (Seeheim, West Germany, 1984), pp. 275-281.
4. J. B. Flanz, "Future Accelerators for Intermediate Energy Nuclear Physics," in Proceedings of the Fifth Course of the International School of Intermediate Energy Nuclear Physics, edited by Roland Bergere, Sergio Costa, and Carlo Schaerf (World Scientific, London, 1986), pp. 287-326.
5. S. P. Kapitza and V. N. Melekhin, The Microtron, edited by E. M. Rowe (Harwood, London, 1978).
6. A National CW GeV Electron Microtron Laboratory, Argonne National Laboratory, December 1982.

7. K. A. Belovintsev et al., "High Current and High Duty Cycle 4.5 GeV Electron Accelerator," Preprint 88, Lebedev Physical Institute, Moscow (1984).
8. H. Herminghaus and H. Euteneuer, "Beam Blowup in Race Track Microtrons," Nucl. Instrum. Methods **163**, 299-308 (1979).
9. "Nuclear Physics Research With a 450 MeV Cascade Microtron," March 1986 report of Nuclear Physics Laboratory, University of Illinois at Urbana-Champaign (unpublished).
10. H. Herminghaus (private communication).
11. H. Euteneuer (private communication).
12. G. A. Loew, "Properties of a Linac-Storage Ring Stretcher System," in Proceedings: Conference on Future Possibilities for Electron Accelerators, edited by J. S. McCarthy and R. R. Whitney (University of Virginia, 1979), pp. Y-1 to Y-27.
13. C. Pellegrini, "Single Beam Coherent Instabilities in Circular Accelerators and Storage Rings," in Physics of High Energy Particle Accelerators (Fermilab Summer School, 1981), edited by R. A. Carrigan, F. R. Huson, and M. Month, AIP Conf. Proc. **87**, 77-146 (1982).
14. R. Servranckx (private communication).
15. William H. Bates Linear Accelerator Center, Pulse Stretcher Ring Conceptual Design Report, August 1986.
16. R. Maas et al., "The Amsterdam Pulse Stretcher," IEEE Trans. Nucl. Sci. **NS-32**(5), 2706-2708 (1985).
17. J. Bisognano and G. Krafft, "Multipass Beam Breakup in the CEBAF Superconducting Linac," to be published in proceedings of 1986 Linac Conference, Stanford.
18. R. L. Gluckstern, "Beam Breakup in a Multi-Section Recirculating Linac," to be published in proceedings of 1986 Linac Conference, Stanford.
19. D. Douglas, "Optics of Beam Recirculation in the CEBAF cw Linac," to be published in proceedings of 1986 Linac Conference, Stanford.
20. A. P. Banford, "The Application of Superconductivity to Linear Accelerators," in International Advances in Cryogenic Engineering: Proceedings of the 1964 Cryogenic Engineering Conference, edited by K. D. Timmerhaus, Vol. 10 (M-U), 1965, pp. 80-87.
21. H. A. Schwettman et al., "The Application of Superconductivity to Electron Linear Accelerators," ibid., pp. 88-97.
22. H. Piel, "Recent Progress in RF Superconductivity," IEEE Trans. Nucl. Sci. **NS-32**(5), 3565 (1985).
23. H. A. Grunder and B. K. Hartline, "Superconducting Accelerator Technology," in AIP Conference Proceedings 150: Intersections Between Particle and Nuclear Physics, edited by Donald F. Geesaman (Argonne National Laboratory, 1986), pp. 37-52.
24. T. Weiland, DESY Rept. 82-015 (1982).

25. P. Kneisel, H. Padamsee, R. Sundelin (private communication).
26. H. Piel and R. Romijn, CERN/EF/RF 80-3 (1980).
27. W. Turchinetz, "Physics Technology--Cost of the New Facilities--A World-Wide Round-Up," in <u>Proceedings for the Conference on New Horizons in Electromagnetic Physics</u>, edited by J. V. Noble and R. R. Whitney (University of Virginia, 1983), pp. 238-272.
28. H. Piel (private communication).
29. <u>CEBAF Design Report</u>, May 1986.
30. F. Netter, R. Bergere (private communication).
31. "Perspectives de la Physique Nucleaire Experimentale a l'Horizon 1995," Rapport de la Commission Mixte IN2P3-IRF, June 1986.

This work was supported by the U.S. Department of Energy under Contract DE-AC05-84ER40150.

ELECTRON ACCELERATORS AT THE FRONTIERS OF PARTICLE PHYSICS

Boyce D. McDaniel
F.R. Newman Laboratory of Nuclear Studies
Cornell University
Ithaca, NY 14853

It is a great pleasure for me to take part in this symposium because it honors a long-time friend of mine--Al Hanson. I first became acquainted with him in 1943--43 years ago--when we were both new PhD's who had joined the war effort at Los Alamos. My first remembrance of him was when he was working with the Van de Graff accelerator in the Physics Division to measure nuclear cross sections. In the years since, we have seen each other at intervals and on occasions have been in contact for making exchange of equipment between the accelerator here and our Laboratory at Cornell. I have admired his initiative, productivity and persistence in furthering the work of this Laboratory. He is both an accelerator builder and accelerator user in the best tradition of the early nuclear physicist.

I would like to start my discussion with the evolution of electron accelerators into tools for high energy particle physics. In Table I are shown some of the major landmarks in the evolution of circular electron accelerators. From my perception, the field began with the Kerst invention of the betatron with the first one being built here at the University of Illinois just before the war in 1941. This was a 2 MeV machine which was followed by the construction of a 22 MeV machine that ended up at Los Alamos. Then just after the war an 80 MeV betatron was built which served as the prototype of the 300 MeV machine which followed. The latter was the last of the big betatrons. Such accelerators had a practical limitation in energy because the accelerating voltage was supplied by the magnetic induction flux inside the orbit which required a very large mass of iron to carry the flux.

At about the same time, applying the invention of McMillan and Vecksler, the first of the synchrotrons was built. These machines provided the accelerating field by means of radio-frequency cavities. In the period from 1946-48 a total of four synchrotrons of about 300 MeV were constructed. Three were patterned after the same mold and were constructed at Cornell, MIT, and Purdue. The fourth machine, using a rectangular flux return frame, was built at Berkeley. These machines were used for measurements of the pion resonances, as well as the first electrodynamics experiments to check the formulas for the gamma ray spectrum in brehmstrahlung. The betatrons and synchrotrons were all of the weak focussing type; that is, the orbit stability was maintained by using a dependence of the magnetic field which fell off as an exponential power, n, of the radius, where the value of n lay between the value zero and one. A value of the exponent near two-thirds was commonly used to avoid an instability at the value of n=1/2.

Table I

Evolution of Circular Electron Accelerators

1. Betatron   Magnetic Induction   Kerst 1940
                       U of Ill.   2, 22, 80, 300 MeV
                       General Electric   100 MeV

2. Synchrotron   Phase Stability   McMillan/Vecksler 1945
                       Cornell, MIT, Purdue   300 MeV
                       Berkeley   300 MeV
                       Cal. Tech.   1.5 GeV

3. Alternating Gradient Synch.   Courant, Livingston, Snyder, 1952
                       Cornell   1.2, 2, 12 GeV
                       CEA   6
                       Daresbury   5
                       DESY   7

4. Storage Rings and Colliding Beams   Kerst 1942
                       Electron-electron   O'Neill, Richter 1959
                       ADA electron-positron   Touscheck 1959
                       VEPP-2   700 MeV
                       SPEAR   2.6 GeV
                       DORIS   3-5
                       VEPP-4   7
                       CESR   8
                       PEP   15
                       PETRA   22

   Under Construction:
                       KEK electron-positron   30 GeV
                       LEP electron-positron   50-100
                       HERA electron-proton   800

Following this period was the invention of strong focussing by Courant, Livingston, and Snyder in 1952. This was a very important step in the progress to higher energy because it greatly reduced the cost of the magnet by reducing the required aperture both vertically and horizontally. Robert Wilson at Cornell had already started the construction of a 1.2 GeV electron synchrotron with weak focussing when Courant, Livingston, and Snyder published their paper. Fortunately, the Cornell machine was designed with demountable poles so that, by a simple modification, the poles could be exchanged to provide strong focussing pole tips. Thus the Cornell machine was the first functioning accelerator to employ strong focussing--even before the Brookhaven AGS. Following the Cornell example, similar accelerators were built at CEA, Daresbury, and DESY. Among the more important physics experiments performed at these machines were the study of the higher resonances in the pion, k meson photoproduction, electron scattering and form factors at high energy.

The next major step in energy advance was the invention of the storage ring and colliding beams. Though it had long been recognized that the center of mass energy in interactions could be increased enormously by colliding two beams, the interaction rates anticipated were distressingly small. To have colliding beams with dynamic accelerators appeared quite infeasible because of the small cross sections involved. By utilizing storage rings in which the beams could be stored for hours, while engaging in collisions with each traversal of the orbit, colliding beams became possible. The earliest experiment that I know for the storage of electron beams was made by Kerst at Los Alamos in 1945 when he clamped the 22 MeV betatron to store the beam at constant energy for about a tenth of a second. In about 1958 O'Neill and Richter started the construction of an electron-electron collider. Shortly after, the Italian group at Frascati, inspired by Touschek, started construction of ADA, an electron-positron colliding beam storage ring. The ADA machine was actually in operation before the electron-electron collider at Stanford. These two machines were the pilot models for the many storage rings that were to follow--the CEA "Bypass", SPEAR, DORIS, VEPP2, CESR, PEP, and PETRA. These machines have had an enormous physics yield including the discovery of the PSI/J, the study of the B meson, jets, electrodynamics and two photon physics.

There are now four higher energy electron machines under construction. These are TRISTAN in Japan, LEP at CERN, SLC at SLAC, and HERA at DESY. TRISTAN and LEP are electron-positron storage rings, SLC is a linear electron collider, and HERA is an electron-proton storage ring. It is anticipated that these tools will make very important contributions and discoveries which will certainly include the discovery of the top quark and perhaps the Higgs particle, the determination of the width of the Z^0 resonance and its decay modes, and the study of e-p collisions at high momentum transfer. Of these machines, LEP is the highest energy circular machine. It is designed for initial operation at an energy of 60 GeV per beam with an ultimate capability of more than 100 GeV per beam. This facility is being built in a tunnel deep underground adjacent to the present CERN site in Geneva, Switzerland. The tunnel circumference is 27 km, the world's largest accelerator tunnel. This project is well along and should become operative in 1988.

All electron-positron storage rings are designed with the electrons and positrons circulating in opposite directions in the same guide field. Aside from the beam energy, the most important feature of a machine is the available interaction rate. This capability of the machine is specified by the quantity defined as the luminosity. The luminosity is that number, which when multiplied by the cross section being studied, gives the event rate for that interaction. Its value is given by the product of the square of the

number of particles per bunch and the frequency with which the
bunches collide, divided by the effective cross-sectional area at the
interaction point. Present colliders have luminosities in the range
from a few times 10^{30} to 10^{32} cm^{-2} sec^{-1}. This luminosity provides
interesting events at the rate of a few per hour to a few per day at
energies up to 100 GeV in the center of mass. The inherent
limitation in the luminosity is generally the beam-beam interaction
which produces an instability as the beam currents are increased.
This limit can be raised by providing strong focussing at the
interaction areas. However, in order to take maximum advantage of
such focussing, the bunch length of the colliding beams must also be
made very short, thus requiring large amounts of accelerating
overvoltage. We have heard this morning about the capabilities of
superconducting r.f. cavities. It is planned at LEP to utilize these
cavities to proceed to the second phase of their program in which it
is hoped that high luminosity at energies even as great as 125 GeV
per beam will be possible. Another approach to reaching higher
luminosity is to increase the number of bunches in the orbit. The
electron and positron orbits are distorted in such a way that
collisions occur only at the chosen interaction points. At Cornell
we are using such a method with seven bunches of each type colliding
at only two interaction points. The luminosity is thus just
increased linearly with the number of bunches.

We are all acquainted with the so-called Livingston plot which
shows how the energy of accelerators has increased with time. In
Fig. 1 is shown a Livingston plot of center-of-mass energy vs. time
for both proton and electron machines. We see that indeed the
electron energy available in the center of mass has increased
exponentially with time. However, looking beyond the present
electron accelerators, we are concerned about our capability to go to
yet higher center-of-mass energies.

At the present time proton accelerators which will provide up to
40 TeV energy in the center of mass are being planned for
construction utilizing conventional accelerator technology and making
use of superconducting guide fields. In proton-proton collisions,
because the proton is a composite quark structure, only about one-
sixth to one-tenth of the center-of-mass energy is available to
produce quark-quark interactions. In spite of the small fraction of
total energy available to the quark-quark interaction, it is many
times greater than that energy which we presently know how to produce
in electron-positron annihilation. The maximum energy planned at LEP
is about 100 GeV per beam, or 200 GeV in the center of mass. It is
likely that circular electron storage rings are limited to energies
only somewhat higher than this.

One problem with pushing the energy frontier is that the cross
sections are falling with the inverse square of the energy, hence
ever higher luminosity is required as the energy is pushed upward.
The second problem is the large capital cost for construction and the
cost of electrical power for operation of circular electron

accelerators and storage rings. Relatively simple arguments show that the capital plus 10-year operating cost for electron machines increases approximately as the square of the energy. This steep functional dependence arises primarily because of the synchrotron radiation losses. The accelerating voltage per turn which is required to make up for the radiation losses increases as the fourth power of the electron beam energy and inversely as the bending radius of the guide field. Because the capital and operating costs for the r.f. system can easily dominate the total cost as the energy is raised, it is clear that there is an advantage to increase the machine radius as the energy is increased. This must, of course, be balanced against the resulting increases in the cost of the guide field system and the costs of conventional construction. One finds that when optimized, machine radius varies approximately as the square of the beam energy. It then follows that the voltage per turn also increases as the square of the beam energy. Thus the total cost, i.e., the sum of the two, increases roughly as the square of the beam energy.

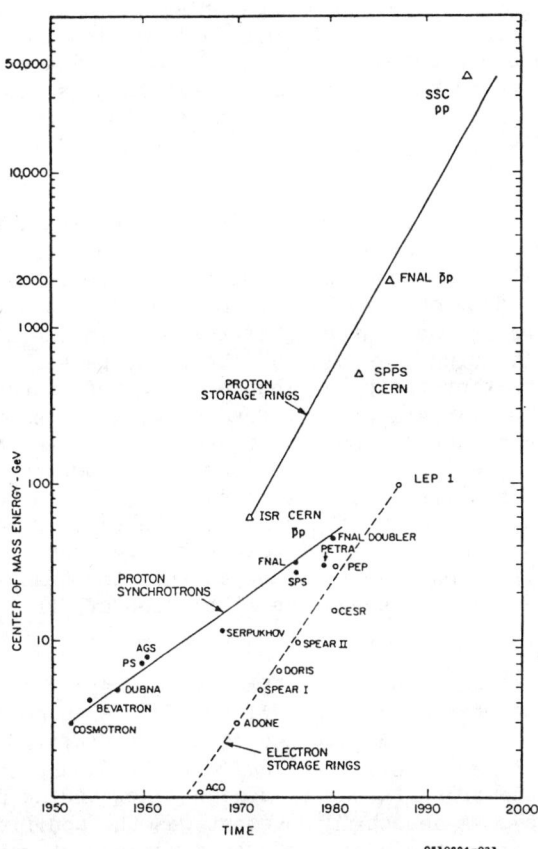

Figure 1: Livingston plot of center-of-mass energy vs. time.

Because of this rapid increase in cost with energy for circular machines, one is led to consider the feasibility of constructing a colliding linac facility. There is only one high energy linear electron accelerator in the world and that is the one at the Stanford Linear Accelerator Center (SLAC). The construction of this accelerator was led by W.K.H. Panofsky and was brought into operation in 1966. It was designed as a 20 GeV linear accelerator working at a wavelength of 10 cm. The energy of this machine has been gradually increased by the addition of more r.f. power until it is currently producing an energy of 35 GeV and is being further improved to reach more than 50 GeV. This linac has been the workhorse for much of the high energy electron physics of the world. In addition to providing beams for very precise measurements in electron scattering, electroproduction, and photoproduction, it has also served as the injector both for SPEAR and PEP colliding beam storage rings.

The linear accelerator has the great advantage of being able to produce very intense beams. Pulse intensities as great as 5×10^{10} can be obtained. It has the disadvantage of having a very low duty cycle. SLAC conventionally operates with a pulse length of the order of one microsecond and a repetition rate limited to 180 Hz, thus the duty cycle is less than one in a thousand. While this low duty cycle is not generally a problem in working with detector systems consisting of narrow aperture spectrometers, it is a disadvantage when making coincidence experiments.

Because of the high pulse beam currents in the linac, an instability called the "head-tail" effect becomes evident. This is an interaction between the beam pulse and the accelerating waveguide as the pulse passes down through the linac. The effect arises because small errors in the centering of the beam induce asymmetric fields in the accelerating waveguide as the pulse passes. The particles in the front of the pulse induce a flow of charge in the walls to produce a wake field which interacts with the particles at the back of the bunch as they arrive. These fields have the effect of forcing the particles at the rear of the bunch further off the center line. The effect is regenerative so that at high beam currents, if nothing is done to overcome it, the tail of the beam pulse may be lost. By making use of frequent, high sensitivity beam position detectors, one may use feedback to keep the beam in the center of the waveguide and reduce the likelihood of difficulty due to this cause.

A conventional linac such as the SLAC linac has a very high power consumption. For operation at its full duty cycle as designed, the power requirement is tens of megawatts. Since its initial installation 20 years ago, several things have been done to upgrade its performance. One of these was to design a new 600 kW power tube of higher efficiency. A second improvement was the modification of the r.f. power system by the addition of r.f. energy storage cavities and switching mechanisms which provide an increase in power to the accelerating waveguide through the superpositon of the r.f. energy of different sections of a long pulse into an r.f. pulse of reduced length and higher power. This mechanism has been called "sledding". It has increased the beam energy by a factor of about 1.4 for the same power input and has permitted SLAC to operate up to energies of 35 GeV.

The linac has demonstrated itself to be very flexible. It is quite easy to direct the beam into various channels with different energies on successive pulses. This interlacing of beam pulses has been very useful to the user community by allowing several different experiments to operate simultaneously. Of course, this does not increase the total charge out of the linac but under appropriate conditions does make a very significant increase in the effective utilization of the machine.

A few years ago, Burton Richter, the present Director of SLAC, proposed that by using colliding linacs a reasonable luminosity for electron-positron annihilation could be attained at energies beyond those practically attainable by the use of storage rings. This was a concept which earlier had not been seriously considered because it was supposed that useful luminosity could not be obtained in this way. However, Richter felt that by pushing several different technological limits, a useful collider could be built. To pursue this concept, he proposed the construction of the Stanford Linear Collider (SLC). This machine would have two functions. The first would be to demonstrate the principle and explore the technological domain of linear colliders, and the other would be to reach an energy adequate to study the production and decay of the neutral boson, Z^0. The center-of mass-energy for production of this particle is 93 GeV. The SLC was approved for construction and it is now about to be commissioned. It is anticipated that the first beam experiments will be conducted starting in January, 1987.

The layout of the collider system is shown in Fig. 2. In the simplest terms, a positron bunch and an electron bunch are accelerated in close spacing down the linac in opposite phases with respect to the r.f. wave. At the end of the linac the two bunches are separated by a transverse magnetic field and are each transported through separate semicircular arcs to come into collision at an interaction region. After collision, the beams are dumped and the cycle initiated again.

In order to obtain high luminosity, the charge bunches must be very compact and contain a large amount of charge. There are two damping rings near the head of the linac which are used to reduce the bunch emittance. Two bunches of positrons are stored in one ring and two bunches of electrons in the other. Within the same pulse of the main linac, one bunch of positrons and one bunch of electrons are injected and accelerated down the full length of the linac to interact at the interaction region. In order to prepare charge bunches for future cycles, during the same linac pulse, the second electron bunch is drawn from the damping ring and accelerated two-

thirds of the way down the linac and deflected onto a converter
target to produce a pulse of positrons which is then returned, after
some acceleration, to be injected into the positron storage ring.
This positron bunch is damped for two cycles before injection into
the linac for acceleration to the collision point. Just after each
pulse of the main linac, two bunches of electrons are injected into
the electron damping ring ready for injection into the linac on the
following cycle. With this cycle of events, collisions can be
obtained on each pulse of the linac.

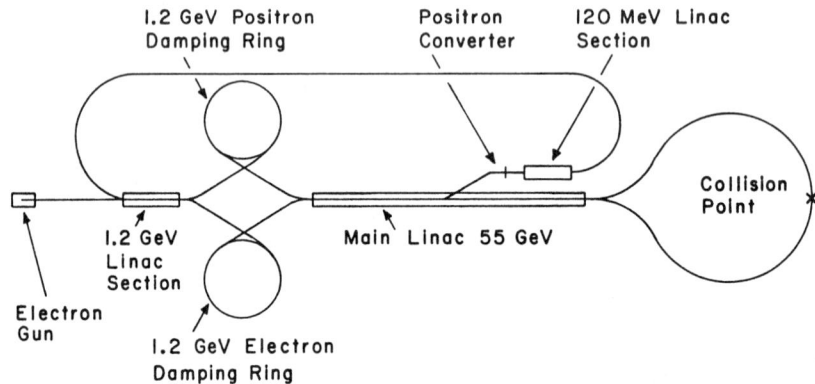

Figure 2

Several very difficult technologies must be mastered in order to
make such a system work. The need for high luminosity imposes a
number of requirements. It is obvious that the luminosity will be
proportional to the repetiton rate, and it is essential that the
intensity of the beam bunches be as high as possible consistent with
the required very small beam size at the interaction area. The
damping rings are designed to provide the low emittance of the beam,
and great care must be taken to maintain the quality of the emittance
throughout the whole accelerator, arc transport, and final focus at
the interaction point. Very strong focussing must be provided in the
early stages of the linac. Because of the high charge in the
bunches, the avoidance of the head-tail instability is essential in
order to maintain the spot size at the collision point. The spot
size in the SLC is supposed to be about 2 microns, hence the beam
position must be controlled with extreme care. In order to make the
beams collide with this precision, it is necessary that there be very
high quality regulation of the bending and focussing magnets in the
transport systems. Even the ground motion must be taken into

consideration. A powerful feedback system is required to keep the beams in collision.

One of the very important factors affecting the luminosity for the colliding linac beams which utilize very high charge bunches is the "disruption" factor, i.e., the impact of one beam bunch upon the other during the time of the collision. Because of the high charge density in the pulse, the bunches focus each other, producing a yet higher charge density during the collision. It is easy to see that such an effect has the right sign when you think of the positron current as a current of electrons going in the opposite direction, that is, in the same direction as the actual electron beam. This gives rise to a magnetic pinch effect without the like-sign electric charge repulsion. The effect of this is to produce a very strong magnetic focussing which lowers the effective interaction cross section of the two beams. This leads to a significant enhancement of the luminosity. If the effect is too strong, it may even blow the beam apart before one bunch passes through the other one. Because this is an effect that depends on the magnitude of the charge in the bunch, the luminosity increases more rapidly than quadratically with the charge. The SLC is designed for an ultimate enhancement factor of 3 to 5.

BEYOND THE SLC

Though the high energy physics community has chosen as its highest long-range priority the Superconducting Super Collider, SSC (a proton-proton storage ring with 40 GeV in the center of mass), this choice is dictated in part by our inability to obtain adequate electron-positron center-of-mass energies with the available electron accelerator technology. However, even if the SSC is constructed, there remains a demanding need for an electron-positron collider to reach or exceed the physics reach of the SSC. This would require obtaining 5 to 10 TeV in the center of mass in electron-positron collisions. The need for such a complementary facility arises largely for two reasons. Because of the very large hadron-hadron background cross sections at the SSC, the difficulty of uniquely analyzing a proton-proton event at the required high luminosity is very great and may make analysis of certain types of events impossible, whereas the backgrounds with electron-positrons collisions at comparable energies are orders of magnitude lower. Furthermore, the electron-positron collider has a simple and uniquely defined initial state which has significant advantages for the understanding of certain physical processes.

Unfortunately, because of the general inverse square cross section dependence of interesting reactions, the demand placed on an accelerator for high energies is very extreme; that is, a very high luminosity is required. For an electron machine this leads to very difficult accelerator design and performance characteristics which are several orders of magnitude better than present capabilities. We can divide these problems into four areas: 1) beam optics, 2) emittance control, 3) power sources, and 4) accelerator structures.

1) <u>Beam Optics</u>. In order to obtain high luminosity at low average beam power, it is necessary to keep the beam spot area small. But the beam spot area is proportional to the product of the beam emittance, ϵ, and the focussing strength, β, at the interaction point. The parameter, β, defines the "depth of focus" or the length along the axis to the point where the spot has increased in area by a factor of two.

Conventionally, the focussing is supplied by a sequence of quadrupole magnets. Individual quadrupoles focus in one plane but defocus in the plane at right angles. By using quadrupole sequences with alternating planes of focus it is possible to focus to a small spot. If beams of very low emittance are provided, then the required aperture near the final focus can be made quite small so that conventional quadrupoles with extremely small gaps may be used to provide the required small β value at the interaction point. Iron quadrupoles with a gap on the order of 100 microns can be imagined. Another alternative would be to use small superconducting quadrupoles which would supply four times the field gradient and would allow a much larger aperture. A third type of quadrupole, the radio-frequency quadrupole, might also be used to provide strong focussing. This device depends on strong electromagnetic r.f. fields in a precisely shaped quadrupole waveguide structure to produce the focussing field. It is visualized that the r.f. field would be produced by a free electron laser.

A fourth possibility, which is even more novel, would utilize the accelerated beam itself to do the focussing. In this case, each beam would consist of two charged bunches closely following each other. The second bunch of each beam would be focussed by the disruption produced by the first bunch of the opposing beam. Thus when the second bunches of each beam collide, they would have already been strongly focussed by the magnetic fields in the first bunches. If one has normal quadrupole focussing with a β of one centimeter, a final focus with a β of about 1 mm could be obtained. These several proposals are well beyond the current state of the art and would require an extended development program before being shown to be practical.

Because of the very small beam spot size, new technologies will be required to control and monitor the alignment of the beam spot which will be a very small fraction of a micron in diameter. The beamstrahlung produced at the interaction point will provide a powerful monitor of the collision and permit feedback of position information to beam steering magnets.

2) <u>Low Emittance Sources and Beam Dynamics</u>. Because the small spot size can only be achieved by the use of a low emittance source, very special attention must be given to this question. The required

emittance for a high luminosity collider is about three orders of
magnitude smaller than the best we know how to produce at the present
time. The only practical way we know to produce very low emittance
beams is by the same method being used in the SLC collider, that is,
by providing a damping storage ring in which the emittance of the
source (particularly the positron source) is reduced by making use of
synchrotron radiation damping prior to injection into the linac. A
certain storage time is required for the injected beam pulse to come
to its reduced emittance. The magnitude of the emittance is
determined by the equilibrium between two factors. The average
radiated energy loss tends to reduce the emittance, but the
fluctuations in that loss tend to increase the emittance. The degree
of damping is determined by the arrangement and strengths of the
guide field magnets and the energy of the beam. The best designs
which have thus far been developed, such as those needed for high
brightness synchrotron radiation sources, are still orders of
magnitude short of the required emittance. The storage times to reach
equilibrium are also uncomfortably long in order to provide damped
beam pulses for a high repetition rate linear collider. Although the
theoretical limit for the emittance far exceeds the requirements,
there are many practical reasons why it will be difficult to reach
the necessary small emittance.

It is not alone sufficent to produce the required emittance; it
is necessary to maintain it throughout the acceleration process. As
described earlier, there is the "head-tail" instability of the beam
which is self-induced as a result of misalignment of the center of
the beam with respect to the axis of the accelerating structure.
Because of this misalignment, the front part of the beam pulse
produces a transverse wakefield in the accelerator so that the tail
of the beam pulse is driven off center. This effect is progressive
as the bunch proceeds down the length of the accelerator. Though the
effect is reduced by reducing the length of the beam bunch, the
situation is much aggravated because of the severe emittance
requirements so that even very small position errors will cause a
significant loss of emittance. The only solution to this problem is
very good control of the beam location during the transit down the
accelerator. This requires very sensitive beam position locaters and
a strong feedback system. Ground motion affecting the alignment of
the accelerator structure and the focus elements also becomes a
serious problem.

3) <u>Power Sources</u>. The type of power source that is to be used
in a high energy linac will depend very much on the wavelength that
is used. For the longer wavelengths the conventional klystron type
of power source appears to be capable of further improvement in order
to provide the very high powers that are needed to operate
conventional cavities; however, because of the question of economy of
operation, much effort is needed to improve the efficiency of such
devices. The peak power capability of this type of source can be
multiplied by the technique of r.f. compression. The "sledding" of
the SLAC linac is one particular system of accomplishing this. In

principle r.f. pulse compression can be applied in multiple stages to give a very large gain in power output in short pulses at the cost of average power.

For overall optimization of a collider, it appears that a wavelength in the range from one centimeter to one millimeter is likely to be the best choice. The free electron laser holds much promise for the generation of very high peak power, especially at shorter wavelengths. The free electron laser consists of an electron beam which passes through a spacially alternating magnetic field which produces a wiggle in the beam. This wiggle generates synchrotron radiation emerging in the forward direction. If a sufficient number of poles are provided and the period of field oscillation is sufficiently short, intense coherent nearly monochromatic radiation may be emitted. This radiation may be stimulated by illuminating the electron beam by a strong electromagnetic pulse of the matching frequency. In effect, the resultant system is a power amplifier of high gain for the input stimulating signal. Such lasers have already been built which have demonstrated an instantaneous power output of 80 MW at a wavelength of about 1 mm. A power gain of 3000 was observed. One of the most interesting experiments currently in progress is the construction of the Two Beam Accelerator at the Livermore Laboratory. This accelerator consists of a 1 cm conventional waveguide which is to be driven at several points by the output of the free electron laser. A low energy, high current (kiloamp) electron beam is used in the laser and this beam runs in parallel with the accelerating waveguide for the driven beam. As the electron beam in the laser passes down the length of the accelerator, energy will be extracted from it because its energy is radiated to drive the accelerator structure. The energy in the driving beam can then be restored by pulsed magnetic induction devices. In essence, the whole system is a transformer in which a high current, low voltage beam is transformed into a low current high voltage beam. This appears to be a promising technology; however, much work remains to be done.

4) <u>Accelerator Structures</u>. Accelerator structures designed for high energy linacs should have the capability of operating at very high gradients in order to minimize the length of the accelerator and hence the capital cost of the structure and its enclosure. As has been remarked earlier, it appears that optimization requires going to a short wavelength. It seems that at 1 cm wavelength, cavities will sustain a gradient up to 500 MeV per meter without breakdown. This means that the power density in the structure is very high, hence the cooling becomes an important factor. Furthermore, because of the very small dimensions of the cavity structure, the accuracy of fabrication and alignment become very important. Another idea which is being explored is the use of open, grating type structures at submillimeter wavelengths which make use of a laser beam passing across the grating to produce accelerating fields. While in principle exceedingly high gradients may be produced in this way, it may be difficult to design a practical device. The use of

superconducting accelerating structures for high energy accelerators holds some promise, but major advances in the achievable gradients and Q are required to make such devices feasible. These improvements do not appear to be limited by the basic characteristics of superconductivity; however, their achievement will require much research and development.

Because there is clearly a great need to be able to exceed LEP energies with electron collisions, a number of accelerator groups in the world are looking seriously at the possibilities for building very high energy linear colliders which are designed around conventional room temperature copper accelerator waveguides. In doing this, however, the various parameters are pushed to their extremes. Richter has initiated a major effort to examine the basic laws governing the design of such high energy, high luminosity colliding electron linacs. He has published sets of parameters which might be possible. One of these is shown in Table II. These correspond to a center-of-mass energy of 10 TeV, a luminosity of 10^{34} cm^{-2} sec^{-1} and the β value of one millimeter at the interaction region. It should be noted that the required beam emittance is a factor of more than one thousand smaller than for the SLC; the spot size is about 10 angstroms, also one thousand times smaller than for the SLC.

Table II

Linear Collider Parameters

	SLC	Beyond SLC	Factor
Energy C.M. (TeV)	0.1	10	100
Luminosity (/cm²/s)	6 x 10^{30}	10^{34}	1600
Spot Radius (micron)	1.5	1.1 x 10^{-3}	1/1300
Emittance (meter)	3 x 10^{-5}	1.2 x 10^{-8}	1/2500
β (mm)	5	1	1/5
Bunch Length (mm)	1.5	1.0 x 10^{-3}	1/1500
Beamstrahlung (%)	0.04	10	250
Beam Power (MW)	0.16	3	18
Rep. Rate (Hz)	180	9000	50

The disruption parameter assumed for the collider is 0.1 which implies an enhancement factor near unity; however, accompanying the disruption factor is a less desirable effect called beamstrahlung. As the bunches pass through each other, their trajectories are sharply bent and the charges radiate synchrotron radiation in the magnetic field. As a result, when the later particles of the bunches collide, the energy spread of the beam is effectively increased. In the SLC, at the maximum disruption, the magnetic fields are of the order of a megagauss but the beamstrahlung spread is only expected to be 0.04 percent. However, in the higher energy machine, even at a disruption factor of 0.1, the beamstrahlung spread is 10 percent, about the maximum which one is willing to tolerate from experimental considerations.

Not apparent in the table is the total A.C. power to be supplied to the system. Current linac designs have very low "wall-plug power" efficiency, delivering only a few percent of the energy to the beam. As a result, without very major improvements in efficiency of acceleration, it is very easy to end up with a requirement for gigawatts of wall-plug power.

I have thus far focussed on rather conventional acceleration technology based on an extrapolation of current methods of acceleration. In recent years, in recognition of the great difficulties that lie ahead to produce very high energy colliding beams, a large number of physicists have begun to create and examine novel ideas which would lead to a reduction in construction costs or operating power for high energy accelerators. It is rather clear that economics is really the governing factor since, in principle, extensions of current technology would, for the most part, permit attainment of very high energies, although at great expense and with apparatus of ridiculous dimensions.

Some novel types of accelerators are described by the following names: plasma, wakefield, switched power, and inverse Cerenkov effect accelerators. Of these, only the plasma and wakefield experiments have progressed to the experimental stage. The plasma accelerator is capable of producing very high acceleration gradients and a pilot experiment has achieved a gradient of about 1000 MeV/meter, but only over a length of millimeters. Such a device has many boundary conditions and the achievement of a practical accelerator is very uncertain.

I will describe an experimental wakefield accelerator which is currently being constructed at the DESY Laboratory in Hamburg, Germany. In this case, a cylindrically shaped low energy high current beam is made to run in a waveguide accelerating structure concentrically with the central high energy beam which is to be accelerated. The structure consists of diaphragms with a central hole for the beam to be accelerated and an annular opening at the outer radius for the cylindrically shaped driving beam. The whole system is enclosed in a conducting outer pipe as shown in Fig. 3. The cylindrically shaped driving pulse is produced photoelectrically by a laser system. This ring of charge is accelerated through the annular opening. As it passes by the diaphragms, fields are produced between the diaphragms which travel inwardly to the central iris. Because of geometrical factors, there is a large multiplication of the field at the central gap, so that the charge bunch receives a strong acceleration as it passes synchronously with the driving charge. In this way, the driving beam loses energy and transfers it to the driven beam. It is anticipated that a transformer ratio of as much as 20 might be obtained producing a gradient of more than 100

MeV per meter. While it appears that one may get a rather high efficiency and high gradient with such a device, there are questions about the symmetry and alignment of the driving beam which may lead to serious instabilities. There is, of course, also the necessity of replenishing the energy of the driving beam, or adding successive stages. I describe this concept to give you a view of the nature of some of these ideas. The pilot model of this particular scheme will be ready to test within a few months.

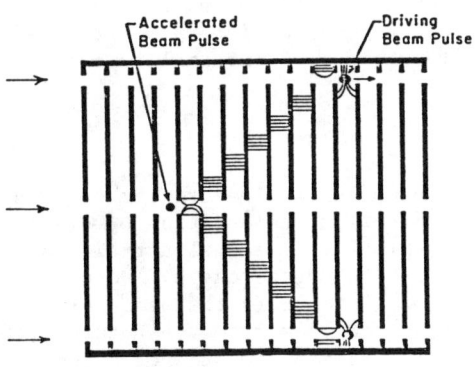

Figure 3: The DESY wakefield transformer.

In summary, during the past 45 years we have seen enormous strides in our capability to construct high energy electron accelerators. We have three new and exciting facilities, LEP, SLC, and HERA, which will become operative in the very near future. However, these machines all fall short in energy when compared to the interaction capability of a large proton-proton collider like the SSC. There are many very novel and entrancing concepts to explore, with the goal of pushing to those energies which are required to complement the physics of the hadron-hadron collider.

Accelerator physics is an exciting and challenging field and is becoming ever more important. I take this opportunity to plead that our educational institutions take a more aggressive role in training students for this profession, and I would encourage young researchers to respond to the challenge.

ACKNOWLEDGMENT

In preparing this paper, extensive use has been made of the Department of Energy document DOE/ER 0255, "Report of the HEPAP Subpanel on Advanced Accelerator R & D and the SSC. December 12, 1985." The author is also much indebted to Dr. Burton Richter, Director of the Stanford Linear Accelerator Center, for providing various documents and transparencies relative to the work at SLAC.

NUCLEAR PHYSICS WITH INTERNAL TARGETS IN ELECTRON STORAGE RINGS

Roy J. Holt
Argonne National Laboratory, Argonne, IL 60439-4843

ABSTRACT

Two key experiments in nuclear physics will be discussed in order to illustrate the advantages of the internal target method and demonstrate the power of polarization techniques in electron scattering studies. The progress of internal target experiments will be discussed and the technology of internal polarized target development will be reviewed.

INTRODUCTION

It is both an honor and a pleasure for me to participate in this thirty-fifth anniversary celebration of electron scattering from nuclei. A perusal of the seminal article[1] by Lyman, Hanson and Scott and the presentations yesterday by Professors Kerst and Hanson indicated that a major obstacle to the early experiments was the extraction of the electron beam from the betatron, so it is ironic that today I shall discuss experiments performed inside an accelerator.

The internal target method is essential for applications which require thin targets, for example, rare targets such as polarized gases or vapors or targets with special isotopic content, and experiments requiring the detection of massive recoiling particles such as electrofission or π^0 production where the recoil nucleus is detected. An indication of the growing interest in the use of internal targets in electron accelerators is summarized[2-6] in the workshops and proposals for electron rings during the past five years (see Refs. 2-6). In addition, there is work[7-8] in progress at Novosibirsk, SLAC and Frascati.

The feasibility of operating an internal target in an electron storage ring is demonstrated by recent work[7] at Novosibirsk where a ^{16}O target in the form of a steam jet was placed in the VEPP-I ring as illustrated schematically in Fig. 1. Here, an average luminosity of 3×10^{31} cm^{-2} s^{-1} at an electron energy of 130 MeV was achieved. Note that the target region was surrounded by NaI crystals which intercepted a total solid angle of 0.6 sr! This is possible only in a high-duty factor and low-background environment. A typical $^{16}O(e,e'\alpha_0)^{12}C$ spectrum from this technique is shown in Fig. 2. The use of the ultra-thin target in this case permitted recoiling α-particles with energies as low as 2.4 MeV to be detected and underscores the usefulness of the internal target technique for detecting heavy recoil particles emerging from the target.

Today, I shall discuss two types of experiments, not only as models for internal target experiments, but also to demonstrate the power of the polarization technique in electron scattering. These two experiments are designed to isolate the charge and quadrupole

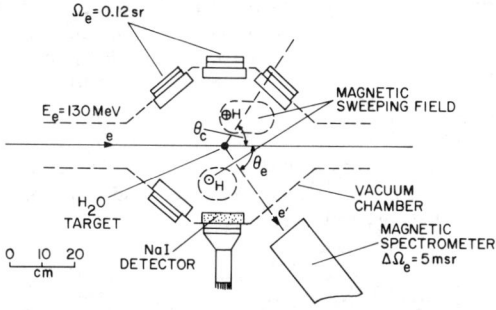

Fig. 1. Schematic diagram of the experimental apparatus at Novosibirsk to perform $A(e,e'x)$ measurements in an internal target geometry. Note the total solid angle of the hadron detectors is 0.6 sr and the average electron beam current is 300 mA.

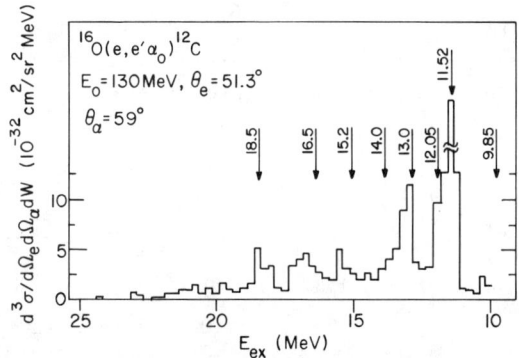

Fig. 2. Typical spectrum for the $^{16}O(e,e'\alpha_0)^{12}C$ reaction. The internal target geometry permits detection of the α-particle at energies as low as 2.4 MeV.

form factors of the deuteron and determine the charge form factor of the neutron, respectively. It will be shown that the central problem with the internal polarized target method is the low target thickness presently available and the development of techniques to improve significantly the target thickness will be discussed.

ELECTRON-DEUTERON ELASTIC SCATTERING

First we consider electron elastic scattering from the deuteron, where three form factors – charge G_0, quadrupole G_2 and magnetic G_1 – describe the scattering process. The primary constraint one can place on present theoretical calculations is to measure the location of the first zero in the charge form factor, which is very sensitive to the model. Indeed, there is great sensitivity to the presence of an isoscalar meson exchange current[9] as well as to quarks in the form of a hybrid model[10] of the deuteron. The location of the first zero in G_0 can only be determined from a polarization measurement. This is evident by noting the form of the unpolarized cross section and the expression for the tensor polarization t_{20} given below.

$$\frac{d\sigma}{d\Omega} = \sigma_M [A(Q^2) + B(Q^2) \tan^2(\theta/2)] ,$$

$$t_{20} = -\sqrt{2} \frac{x(x+2) + y/2}{1 + 2(x^2+y)}$$

where x depends on the ratio of the quadrupole to the charge form factor,

$$x = \frac{2}{3} \tau G_2/G_0 ,$$

and y depends on the square of the ratio of the magnetic to the charge form factor and is very small below $2(GeV/c)^2$,

$$y = \frac{2}{3} \tau (G_1/G_0)^2 f(\theta) ,$$

where

$$f(\theta) = \frac{1}{2} + (1+\tau) \tan^2(\frac{\theta}{2}) ,$$

and

$$\tau = Q^2/4M_d^2 .$$

Clearly, only A and B can be isolated in a measurement of the cross section, and it turns out that the first zero in G_0 is masked by the presence of G_2 in the quantity A. The quantity $B(Q^2)$ which depends only on G_1 has been well determined[11,12] up to a momentum transfer of $2.5(GeV/c)^2$ and it is found that G_1 has little influence on the value of t_{20} in this momentum transfer range. Then, t_{20} is essentially providing information on the ratio of G_2/G_0 and, in fact, as $G_0 \to 0$, $t_{20} \to -\frac{1}{\sqrt{2}}$ in this approximation that G_1 is negligible. Thus, a measurement of the quantity t_{20} provides the best information on the location of the first zero in G_0.

Although there has been much previous speculation[13,14] that perhaps the effects of perturbative QCD might be observed in t_{20} in this momentum transfer region, the recent measurements[12] of $B(Q^2)$ at SLAC indicate that an explanation of the deuteron in terms of hadrons appears to be predominant up to a momentum transfer of $2.5(GeV/c)^2$.

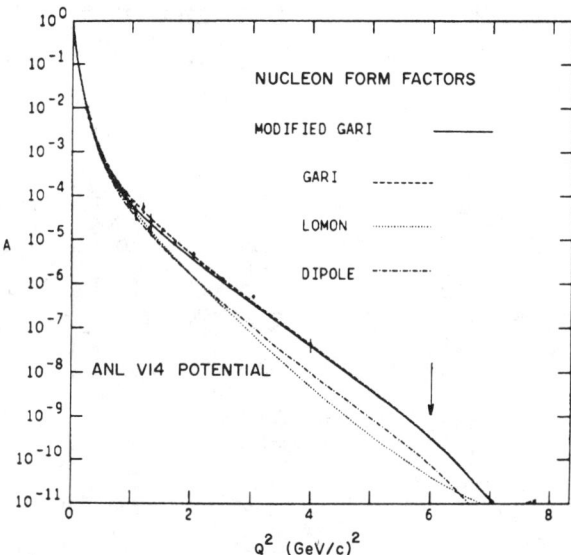

Fig. 3. The quantity $A(Q^2)$ as a function of Q^2. The data are from Ref. 21, and the calculations are performed for three different nucleon form factors. Above a momentum transfer of 1 $(GeV/c)^2$ the choice of the nucleon form factor is critical to the interpretation of the data.

Another issue which is re-emerging is the uncertainty in the nucleon form factors. A recent analysis[15] of cross section data for the nucleon has indicated that the charge form factor of the neutron is much larger than previously believed. The effect of this present uncertainty on the quantity $A(Q^2)$ is illustrated in Fig. 3 where recent calculations by Chung et al.[16] are provided kindly by F. Coester for this 35th anniversary celebration. These calculations employ the Argonne V14 potential[17] for the deuteron wave function and the relativistic front-form dynamics[18] which has recently been applied to electromagnetic interactions by Coester. There is a rather large discrepancy between the recent form factors by Gari[15] and the more conventional[19] form factors. Note that the choice of the Gari form factors would rule out a correction from the isoscalar meson exchange current of the type discussed by Gari and Hyuga.[20] It is clear from this graph that more work on the nucleon form factors is necessary in order to advance our understanding of the deuteron. In a few moments I will discuss the role that polarization studies will have in improving measurements of the nucleon form factors.

The tensor polarization in electron deuteron scattering, however, has little dependence on the choice of the nucleon form factors owing to the fact that it is primarily dependent on the ratio of the electric form factors of the deuteron. Calculations of the tensor polarization, given in Fig. 4, exhibit a strong

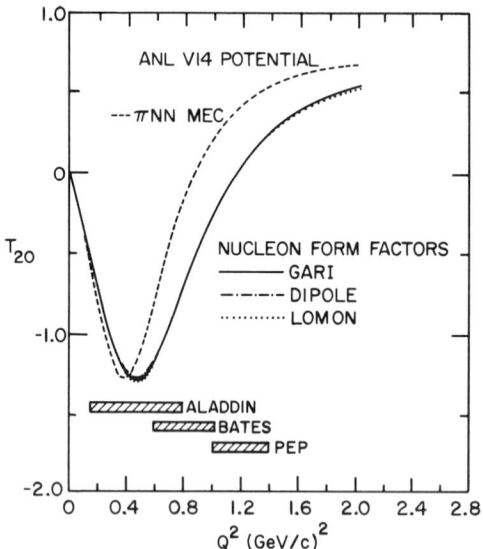

Fig. 4. Calculations of the tensor polarization both with and without the isoscalar meson exchange current. The effect of employing the different nucleon form factors is negligible for t_{20}. The hatched regions indicate the momentum transfer region for three separate proposed experiments.

dependence on the isoscalar meson exchange current, while simultaneously showing no discernible effect from the choice of the nucleon form factors. In addition, the tensor polarization has a small dependence on the choice of realistic deuteron wave functions. I believe that this represents the most sensitive test available for the isoscalar meson exchange current in nuclei. Of course, there are other calculations involving hybrid, six-quark bag models[10] and deltas in nuclei,[22] which indicate that t_{20} might be sensitive to the presence of quarks or deltas in the deuteron. Unfortunately, these calculations have not achieved the same degree of sophistication as that of the calculations presented here. For example, with regard to detecting the presence of deltas in nuclei, there is the open question of the effect of including the delta-delta interaction and the choice of the form factors for the delta. However, measurements of the polarization which occur significantly outside the region bound by the curves in Fig. 4 would be a signature for new physical processes of this kind. An indication of the momentum transfer region for new measurements of t_{20} are shown by the hatched regions at the lower part of the figure. Proposed internal target experiments[14,23,24] are denoted by the label Aladdin or PEP, which represent existing electron storage rings of energy 1

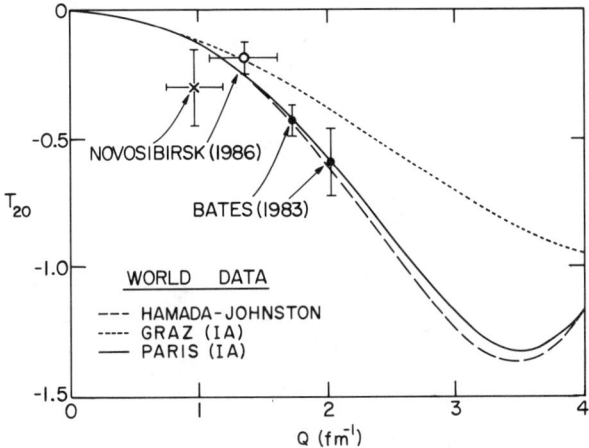

Fig. 5. Summary of all measurements of t_{20}. The two values at the highest momentum transfer were measured in an experiment at MIT-Bates where a polarimeter was employed to measure the polarization of the recoil deuterons. The other two values were measured in an internal target geometry at Novosibirsk.

GeV and 14.5 GeV, respectively. The region labelled Bates represent the momentum transfer range which should be accessible in an experiment proposed[25] at the MIT-Bates Laboratory.

MEASUREMENT OF POLARIZATION IN ELECTRON SCATTERING

Previous measurements[26] of tensor polarization have been performed up to a momentum transfer of 0.16 $(GeV/c)^2$ with the use of a deuteron tensor polarimeter at the MIT-Bates Laboratory and are shown in Fig. 5. Additional measurements of t_{20} are planned at Bates to extend this range between 0.5 and 1.0 $(GeV/c)^2$. Experiments involving deuteron tensor polarimeters are extremely difficult in that it is necessary to perform the calibration of the polarimeter at a separate laboratory and ensure that the same conditions exist at the electron scattering facility. This additional source of systematic error is not present in an internal polarized target geometry. The amount of hardware for the external beam experiment is frequently cumbersome involving a large acceptance magnetic spectrometer in order to detect the electrons, a complicated large acceptance S-bend magnetic channel to direct the deuterons to the polarimeter, massive amounts of shielding and a high-power liquid deuterium target in addition to the calibrated polarimeter.

The advantage that the internal target method holds over an external beam experiment is best illustrated by considering a pilot

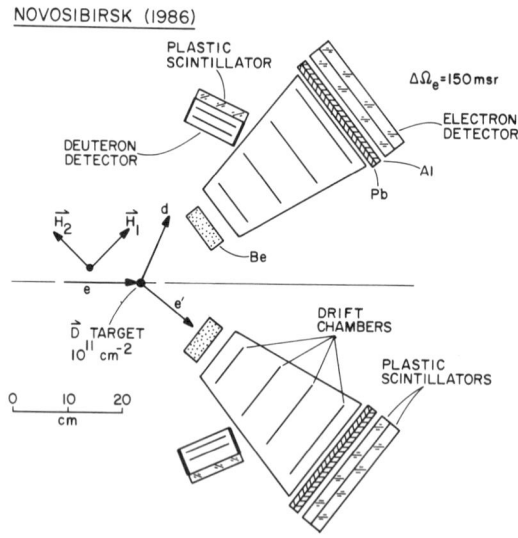

Fig. 6. Schematic diagram of the internal polarized deuterium target experiment at Novosibirsk. Note that large solid angle detectors were employed in the experiment in the presence of a 300 mA electron beam. The target thickness was only 10^{11} nuclei/cm^2.

experiment[27] which was performed at Novosibirsk and is illustrated schematically in Fig. 6. The most instructive lesson to be learned from this experiment is the relative simplicity and small scale of the detector system. Since the background in an internal target geometry is minimal, large solid angle detectors can be employed for both the scattered electrons and deuterons. The use of an average electron current of approximately 0.3 A and along with the large solid angle detectors (150 msr) emphasizes the small background in the internal target geometry. Even with the high current and large solid angle detectors, the experiment was limited by two key factors: small target thickness (10^{11} nuclei/cm^2) and low electron beam energy (400 MeV). A review of the internal target technology will be discussed in a moment, but it is expected that a target thickness of $\gtrsim 10^{14}$ nuclei/cm^2 can be achieved. Thus, the internal target method could, perhaps, represent the most powerful method for the study of polarization in electron-deuteron scattering.

CHARGE FORM FACTOR OF THE NEUTRON

One of the most elusive problems for electron scattering has been the determination of the charge form factor of the neutron.

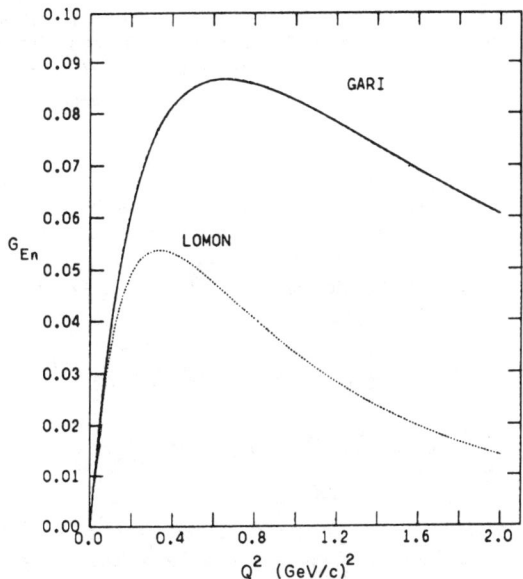

Fig. 7. Results of two analyses of the charge form factor of the neutron.

Earlier I emphasized the importance of measuring the nucleon form factors, and especially that of the neutron, in order to resolve issues in electron-deuteron scattering. Yesterday, Professor Hofstadter presented beautiful results for the form factor of the neutron based upon quasifree electron scattering from the neutron in the deuteron. However, an ambiguity in the analysis of these kinds of data seems to persist even today. As an indication of this ambiguity the charge form factors extracted from two analyses are illustrated in Fig. 7. The most recent analysis by Gari constrains the model to obey vector dominance at low momentum transfer and perturbative QCD at very high momentum transfer. The result is that the Dirac form factor F_{1n} becomes vanishingly small throughout the entire momentum transfer region and leads to a rather large form factor for the neutron. The difficulties of the measurement arise from the fact that there is no "clean" neutron target and that the charge form factor G_E is much smaller than the magnetic form factor G_M. Unlike the case for a spin-one nucleus, a standard Rosenbluth separation can, in principle, resolve the charge and magnetic form factor of the nucleon. However, the charge form factor is obscured since G_E and G_M appear in the cross section as the sum of the squares of these quantities.

This problem can be alleviated greatly with the use of a polarization technique, where it is necessary to utilize a polarized electron beam as well as a polarized target. Consequently, it would be essential to have a dedicated ring for internal target studies of this kind and it is good to hear that proposals for internal target facilities exist at both the MIT-Bates and NIKHEF Laboratories. Fortunately, Norum has discussed[28] a technique for preserving longitudinal polarization of the electrons at an internal target location in a storage ring.

Although this experiment has been discussed[29] by H. Jackson elsewhere, I wish to summarize the essential points here. The expression for the cross section for longitudinally polarized electrons scattering from a polarized nucleon target as indicated in Fig. 8 is given according to Donnelly and Raskin[30] as

$$\sigma(\theta_e, \theta^*, \phi^*) = \sigma_M f_{rec}^{-1} \{V_L(1+\tau^2)G_E^2 + 2V_T\tau(1+\tau)G_M^2 +$$
$$-2 P_e P_T [\tau(1+\tau)V_T' G_M^2 \cos\theta^* - (2\tau)^{1/2}(1+\tau)^{3/2} V_{TL}' G_E G_M \sin\theta^* \cos\phi^*])$$

where the V's, τ and f_{rec} are kinematic factors given by in Ref. 30, P_e and P_T are the electron and nucleon polarization, respectively. The most significant feature of this expression is the last term in which G_E and G_M enter linearly rather than as the sum of the squares and thus a polarization measurement remarkably increases the sensitivity to G_E. The orientation angles θ^* and ϕ^* of the target spin can be changed rapidly and precisely with an internal target geometry since only a magnetic field of several gauss is employed as a guide field; whereas, this process could consume hours for an external polarized target in which a 35 kg field must be reversed. The time required for a measurement of G_E to an accuracy of ± 20% with a luminosity of 10^{33} cm^{-2} s^{-1} and an energy (0.88 GeV) suitable for the proposed MIT-Bates ring is shown in Fig. 9. Note that a somewhat higher energy (1.3 GeV) produces a large reduction in the time required for the experiment. However, an energy higher than 2 GeV does not improve substantially the result, since the electrons for a given momentum transfer are scattered more forward in angle and the resulting asymmetry becomes smaller.

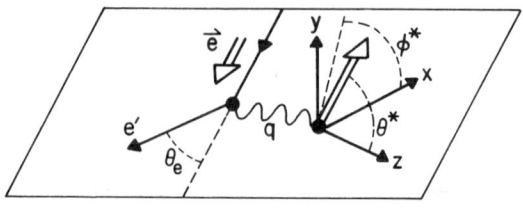

Fig. 8. Orientation of the spins of the electrons and the nucleon target in an electron-nucleon elastic scattering process.

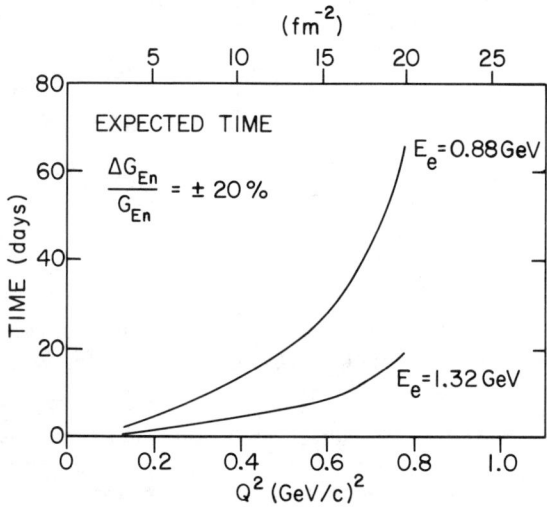

Fig. 9. Time estimate for a typical experiment at the proposed MIT-Bates ring to measure the charge form factor of the neutron to an accuracy of ± 20% and at two electron energies.

The choice of a target for the polarized neutron is critical, and thus far, ^{2}H and ^{3}He have been identified as candidates. Although the deuteron is considered the best case for the study of the neutron form factor, Blankleider and Woloshyn[31] point out that ^{3}He has the advantage that the magnetic scattering from the two protons in ^{3}He is minimized owing to the paired-off spins in the dominant configuration of the ^{3}He wave function. Thus, it is expected that both polarized deuterium and ^{3}He targets will have a major role in these measurements and I would like to turn to a discussion of recent developments in the internal polarized target technology.

DEVELOPMENTS IN INTERNAL POLARIZED TARGET TECHNOLOGY

The most significant progress in polarized gas or vapor targets in recent years has been reported[32,33] by two optical-pumping groups. In particular, it has been instructive to follow the progress of the groups at KEK and TRIUMF where relatively high densities of Na have been optically pumped in order to fabricate a high flux H$^-$ source. The approximate geometry of the KEK and TRIUMF work is illustrated schematically in Fig. 10. The two prominent features of this Na target are that it is windowless and no buffer gas is used in the optical pumping process. The TRIUMF group recent reported[33] a new record in target thickness: a Na↑ target thickness

OPTICALLY PUMPED TARGETS

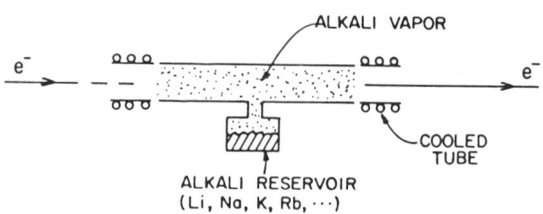

Fig. 10. Illustration of a possible geometry for a polarized internal vapor target.

of nearly 10^{14} nuclei/cm^2 and a polarization of 60% have been achieved. The TRIUMF group reports that no fundamental limitations have been observed as yet and they are planning to double the laser power in order to improve upon these results.

LeDuc et al.[34] have demonstrated that high densities ($\sim 10^{17}$ nuclei/cm^3) and polarization ($\sim 50\%$) of ^{3}He nuclei can be achieved in the laboratory by direct optical pumping of metastable ^{3}He. Work[35] is in progress at Caltech in order to assess the suitability of this technology for producing an internal or external polarized target of ^{3}He. Note that present-day, conventional external polarized ^{3}He targets[36] involve driving ^{3}He into its frozen solid state and heat from even the smallest amount of electron beam (picoamps) would destroy the target polarization. Thus far, the Caltech group have drawn two conclusions: (i) development of an internal polarized ^{3}He target of density $\gtrsim 10^{15}$ nuclei/cm^2 and a polarization $\gtrsim 50\%$ appears to be practical, (ii) a high-density polarized ^{3}He target for use in an external electron beam appears to be practical for polarized electron beam currents exceeding 80 μA. In addition, it appears that recent developments[37] in polarized electron sources, particularly the ionization of optically-pumped metastable ^{4}He atoms, will be beneficial to the new generation of CW electron accelerators and this high current of polarized electrons should be practical.

As a final example of the technology I shall summarize the progress of attempts at Argonne to harness the spin-exchange optical pumping method in order to produce a polarized deuterium target. Here, the goal is to produce a target thickness of 10^{14} nuclei/cm^2 with a tensor polarization $t_{20} \gtrsim 0.3$. In order to achieve this density, it is necessary to utilize the highest flux source of polarized deuterium available as well as retain the atoms in a storage cell, as illustrated in Fig. 11. With the constraints that the storage cell should not exceed 10 cm in length and the atoms should not surpass 1000 collisions with the walls of the storage container before leaking out the "windows", it is imperative that polarized atoms be injected into the storage cell at the rate of 4×10^{17} s^{-1}.

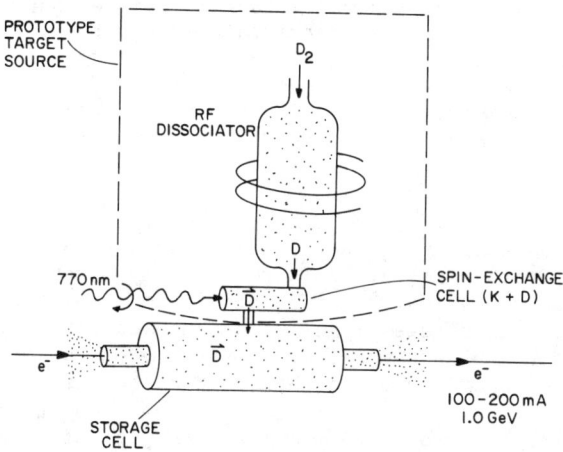

Fig. 11. Illustration of a possible internal polarized deuterium target geometry. The region oulined by the dashed curve denotes the prototype source which is being tested presently at Argonne National Laboratory.

This high flux exceeds the capability of conventional[38] atomic beam sources by an order of magnitude, and, consequently, the Argonne group is investigating the spin-exchange optical pumping method as a means of producing a high flux source. The primary difficulty with the conventional source is associated with the use of a hexapole magnet to polarize the atoms. The highest possible density of atoms is presented to the entrance of the hexapole and this density is limited by atom-atom collisions and recombination of the atoms (deuterium or hydrogen) at the entrance. Unfortunately, the hexapole can only transmit a small fraction of these atoms owing to the relatively small solid angle. In the spin-exchange optical pumping method, the hexapole magnet is eliminated and there exist no high-density regions which remove one from the molecular flow regime. At present, it is believed that the flux from this model source is limited only by laser power which would impose a practical flux of approximately 10^{18} atoms/s. Another important advantage of this technique over that of the conventional source is the relatively low gas load from unwanted atoms, so that the novel source is well-suited for use in an electron storage ring.

Specifically, the process which has been discussed[23,39] previously involves polarization of K atoms by optical pumping and polarization of the deuterium by successive collisions with the K↑ atoms. It is essential that the loss of polarization by atomic collisions with the walls of the spin-exchange cell be minimized. Fortunately, a number of surface coatings have been studied by two groups[40,41] at Wisconsin, and we have found[42,43] a coating known as drifilm to have suitable properties for the spin-exchange method. Presently, a prototype of the polarized source, shown as the outlined region in Fig. 11, is undergoing tests at Argonne. If these tests prove to be positive, the longer range goal is to develop a tensor polarized deuterium target for use in the Aladdin storage ring at Wisconsin, and perhaps, the PEP ring at SLAC.

SUMMARY

It was shown that the hitherto unexploited polarization technique in electron scattering is potentially a very powerful method for nuclear physics. The feasibility of employing internal targets in electron storage rings in conjunction with large solid angle detectors has been demonstrated by the group at Novosibirsk in which $^{16}O(e,e'x)$ experiments and $^{2}\vec{H}(e,e)^{2}H$ measurements were performed recently. Recent progress in optical pumping appears to be very promising for producing polarized targets of practical thicknesses. Finally, I predict that during the next thirty-five years of electron scattering studies, polarization measurements will have an important impact on our understanding of nuclei.

ACKNOWLEDGEMENTS

I wish to acknowledge my collaborators who have participated in the development and testing of the Argonne prototype polarized deuterium source: D. F. Geesaman, M. C. Green, R. S. Kowalczyk, G. E. Thomas, L. Young and B. Zeidman. A special thanks goes to M. Peshkin for providing a theoretical treatment of the spin-exchange optical pumping process. We thank B. Norum for contributions to our understanding of internal targets in storage rings and for assistance with tests of the prototype target. In addition, I wish to thank F. Coester who kindly agreed to allow presentation of recent calculations prior to their publication. Finally, I wish to thank J. Berkowitz and L. Goodman for very useful discussions.

This work supported by the U.S. Department of Energy, Nuclear Physics Division, under contract W-31-109-ENG-38.

REFERENCES

1. E. M. Lyman, A. O. Hanson and M. B. Scott, Phys. Rev. 84, 626 (1951).
2. Proceedings of the Workshop on Nuclear Physics with the Use of Electron Storage Rings, Lund, Oct. 5-7, 1982, University of Lund Report.
3. Proceedings of the Workshop on Polarized Targets in Storage Rings, Argonne, May 17-18, 1884, ANL-84-50.
4. Proceedings of the CEBAF Summer Workshop, Newport News, June 25-29, 1984; June 3-7, 1985.
5. Proposal for the MIT-Bates Pulse Stetcher Rings, MIT, June 8, 1984.
6. Workshop on Polarized Targets: New Techniques and New Physics, Bull. Am. Phys. Soc. 31, 1195 (1986).

7. B. B. Voisehovski et al., preprint, Novosibirsk (1986); S. G. Popov, Proceedings of the Workshop with Electron Rings, for Nuclear Physics Research, Lund, Oct. 5-7, 1984, p. 150.
8. F. Dietrich et al., Proc. of the Second Conf. on the Intersections Between Particle and Nuclear Physics, Lake Louise, May 1986, AIP Conf. Proc. 150, p. 378.
9. M. I. Haftel et al., Phys. Rev. C 22, 1285 (1980).
10. A. P. Kobyshikin, Sov. J. Nucl. Phys. 28, 252 (1978); I. L. Grach and L. A. Kondratyuk, Sov. J. Nucl. Phys. 39, 198 (1984); L. Kisslinger and H. Ito, preprint (1986).
11. S. Auffret et al., Phys. Rev. Lett. 54, 649 (1985) and references therein.
12. P. E. Bosted, Proc. of the Second Conf. on the Intersections Between Particle and Nuclear Physics, Lake Louise, May 1986, AIP Conf. Proc. 150, p. 554.
13. C. Carlson and F. Gross, Phys. Rev. Lett. 52, 1080 (1984); S. J. Brodsky and B. T. Chertok, Phys. Rev. Lett. 53, 127 (1974).
14. R. J. Holt, Proceedings on Intersections Between Particle and Nuclear Physics, AIP Conf. Proceedings, No. 123, 499 (1984).
15. M. Gari and W. Krümpelmann, preprint (1986); M. Gari, preprint (1986).
16. P. L. Chung et al., private communcation (1986).
17. R. B. Wiringa et al., Phys. Rev. C 29, 1207 (1984).
18. F. Coester, Bates Users Theory Group Workshop, MIT, Aug. 9-10, 1985, ANL preprint PHY-4667-TH-85; Workshop on Constraints Theory and Relativistic Dynamics, INFN, Florence, May 1986, ANL preprint PHY-4804-TH-86.
19. E. Lomon, Ann. of Phys. 125, 309 (1980).
20. M. Gari and H. Hyuga, Nucl. Phys. A264, 409 (1976).
21. R. Cramer et al., Z. Phys. C 29, 513 (1985); and references therein.
22. R. Dymarz and F. C. Khanna, Phys. Rev. Lett. 56, 1448 (1986).
23. R. J. Holt, Proceedings of the Workshop on Polarized Targets in Storage Rings, Argonne (1984), ANL Report ANL-84-50, p. 103.
24. R. J. Holt et al., Nucl. Phys. A446, 389c (1985).
25. L. Antonuk et al., MIT-Bates Proposal No. 84-17.
26. M. E. Schulze et al., Phys. Rev. Lett. 52, 597 (1984).
27. V. F. Donitriev et al., Phys. Lett. 157B, 143 (1985); D. K. Vesnovski, preprint 86-75, Novosibirsk (1986).
28. B. Norum, Report of the 1985 CEBAF Summer Study, Newport News, VA, 1985, p. 17-70.
29. H. E. Jackson, ibid. 23, p. 53.
30. T. W. Donnelly and A. S. Raskin, Ann. Phys. 169, 247 (1986).
31. B. Blankleider and R. M. Woloshyn, Phys. Rev. C 29, 538 (1984).
32. Y. Mori et al., Proceedings of the Conf. on Polarized Proton Ion Sources, TRIUMF, Vancouver (1983), AIP Conf. Proc. 117, p. 123, Nucl. Instrum. Meth. 220, 264 (1984).
33. C. D. P. Levy et al., Preprint 1986.
34. M. LeDuc et al., Nucl. Sc. Applications 1, 1 (1983).
35. R. D. McKeown and R. G. Milner, Report of the 1985 CEBAF Summer Study, Newport News, 1985, p. 12-45 (1986).
36. D. G. Haase and C. R. Gould, Bull. Phys. Soc. 31, 1226 (1986).

37. L. G. Gray et al., Rev. Sci Instrum. 54, 271 (1983).
38. H. G. Mathews et al., Nucl. Inst. Meth. 213, 155 (1983); W. Grüebler, ibid. 23, p. 223.
39. M. C. Green, ibid 22, p. 307; Worskhop on Nuclear Physics with Stored, Cooled Beams, AIP Conf. Proc. 128 (1985), p. 268.
40. W. Haeberli, ibid. 38, p. 251.
41. D. R. Swenson and L. W. Anderson, Nucl. Instr. Meth. B12, 157 (1985); L. W. Anderson and D. R. Swenson, private communication (1986).
42. G. E. Thomas et al., Proc. of the Thirteenth World Conf. on Nuclear Target Development, Chalk River, September 1986, to be published.
43. L. Young et al., Proc. of the Ninth Conf. on Applications of Accelerators in Research and Industry, Denton, TX, November 1986, to be published.

> The submitted manuscript has been authored by a contractor of the U. S. Government under contract No. W-31-109-ENG-38. Accordingly, the U. S. Government retains a nonexclusive, royalty-free license to publish or reproduce the published form of this contribution, or allow others to do so, for U. S. Government purposes.

ACCURACY OF THE ELECTRON PROBE

D. G. Ravenhall
University of Illinois at Urbana-Champaign, Urbana, IL 61801

ABSTRACT

An attempt is made to assess the relative accuracy of various approximations used in high-energy electron-nucleus scattering, especially as applied to more exclusive reactions involving polarization of beams or targets, and detection of nuclear final-state polarizations.

INTRODUCTION

When planning experiments with electrons or examining qualitatively their results, we think in terms of the plane wave Born approximation (PWBA). It is appealing because of its simplicity. It is, however, an approximation: as the Feynman diagram in Fig. 1 illustrates, it includes the exchange of only one photon. In this talk I wish to discuss the improvements which are, or can be, made on this approximation, and the changes they produce.

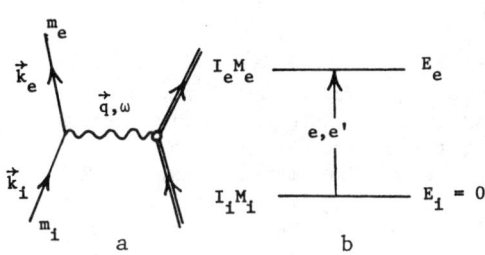

Fig. 1 (a) diagram for plane wave Born approximation to (e,e') and (b) level diagram of nuclear excitation.

This is not a review paper. It was prepared at very short notice and contains therefore only items which were readily accessible to me. It is certainly cursory and probably slanted. I apologise in advance to those people whose relevant work is not mentioned.

It was evident from the first (e,e) experiments that for elastic scattering of electrons by nuclei, the observed filling in of the Born-approximation zeroes of the differential cross section, more noticeable as the nuclear charge Z was increased, implied that there was multiple exchange of photons generated by the nuclear Coulomb field. It rapidly became mandatory to take this into account in analysing elastic scattering, by solving the Dirac equation using a phase shift analysis. The inclusion of these soft photons into the analysis of inelastic scattering took only a little longer, and by now it is performed by most experimenters. That does not exhaust the possible photon exchanges, however. It is possible to have multiple exchanges of hard photons which involve virtual excitation of other nuclear states than the ground and final states. (In these processes the nucleus is treated not as an inert object but, more completely, as a dynamical system.) Such treatments have been applied to only a few experimental situations, however. Thus there are various levels of sophistication possible in treating the problem: PWBA, DWBA and hard-photon, or coupled-channels (as we shall call it).

For the (e,e') excitation of discrete nuclear levels, there are by now various layers of complexity and exclusivity accessible to experiment. To be contrasted with the classic unpolarized (e,e') process, possibilities include:
  orientation of the target nucleus;
  polarization of the incident electron beam;
  detection of the M_e state population of the final
    nuclear state by measuring angular distributions of
    decay products (such as γ rays);
  combinations of the above.
The question of interest, to which we have in most cases only fragmentary answers, is the extent of the multi-photon corrections, at the various levels (DWBA, or coupled channels), on these different exclusive experimental situations. Our impression is that as the processes become more exclusive, multiphoton exchange becomes relatively more important, especially since the details sought by these newer experiments are usually finer, and the precision required is greater.

Let us look at some of the things that have been done.

ELECTRON SCATTERING AND NUCLEAR CHARGE DISTRIBUTIONS

For reference, we recall the results of the Plane Wave Born Approximation (PWBA) for scattering from ground-state or transition charge distributions. The quantities we call the charge densities, $\rho_\ell(r)$, are contained in the expansion of the matrix element of the nuclear charge operator in spherical harmonics:

$$\langle I_e M_e | \rho(\vec{r}) | I_i M_i \rangle = \sum_\ell C^{I_i \ell I_e}_{M_i -mM_e} \frac{1}{\sqrt{(2I_e+1)}} \rho_\ell(r) Y_{\ell m}(\hat{r}) \qquad (1)$$

For unpolarized electrons on unpolarized nuclei, the differential cross section is

$$\frac{d\sigma}{d\Omega} = \left(\frac{2e^2 k_e \cos\theta}{\vec{q}^{\,2}}\right)^2 \frac{1}{2I_e+1} \left(F^C_\ell(q)\right)^2 , \qquad (2a)$$

where F is the form-factor for the ℓ-th multipole,

$$F_\ell^C(q) = \sqrt{4\pi} \int_0^\infty j_\ell(r)\, r^2 dr. \qquad (2b)$$

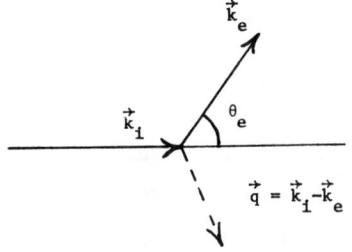

Fig. 2 Definition of $\vec{q}$.

The momentum transfer q is defined in Fig. 2, and by the expressions

$$q = [4k_i k_e \sin^2 \tfrac{1}{2}\theta_e + \omega^2]^{1/2}$$
$$\approx 2k_i \sin \tfrac{1}{2}\theta_e \qquad (3)$$

Distorted Wave Born Approximation

For all except the lowest Z nuclei it is necessary to allow for the distortion of the electron incident and final wave functions by the Coulomb potential generated by the nuclear charge distribution (the $\ell=0$ part of the expansion (1) for the diagonal matrix element in the initial and the final state respectively). This is shown in Fig. 3. As was mentioned in the introduction, this procedure has to be gone through for elastic scattering in order to determine the ground-state $\rho_0(r)$. To understand what is going on (although not to make the actual experimental analysis) one can use an eikonal approximation to these wave functions.[1] Pictures of the two wave functions, showing the wave fronts and rays (normals to the wave fronts) taken from Ref.1 are shown in Fig. 4. For the relatively short wavelengths implied by the illustrations the Coulomb potential $V_C(r)$ appears slowly-varying, and it has the effect of focussing the electron waves. The wave function in the neighborhood of the nucleus has a wave number which is spatially varying, and somewhat larger than its asymptotic value because V_C is attractive. The normalization is altered correspondingly, as shown in Fig. 4b. When wave functions of this kind are used to calculate the scattering amplitude

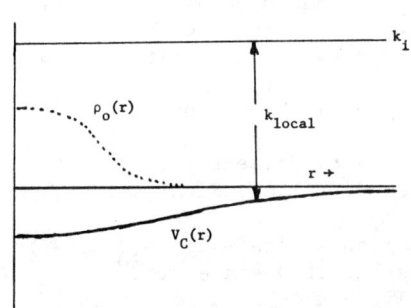

Fig. 3 The Coulomb potential $V_C(r)$.

$$A = \int d^3r\, \psi_e^{(-)*}(\vec{r})\, V_{ex}(\vec{r})\, \psi_i^{(+)}(\vec{r}), \qquad (4)$$

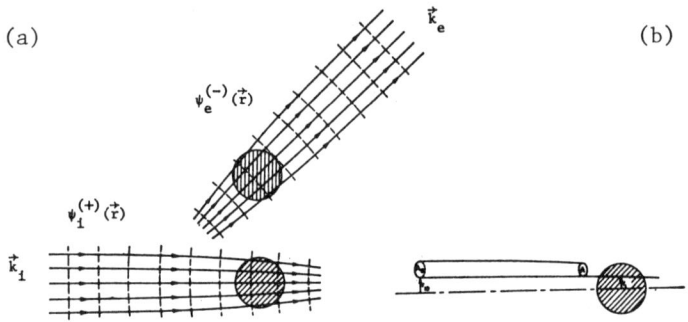

Fig. 4 Left, eikonal wave functions; right, the normalization of the initial wave function.

where $V_{ex}(\vec{r})$ is the Coulomb potential generated by the transition charge density, the resulting differential cross section has the features we expect: the diffraction zeroes are filled in by an amount which depends on the depth of V_C, and the diffraction structure is shifted to smaller q-values because of the modified wave-number (also due to V_C).

Since the start of inelastic scattering measurements, computer programs have been constructed which obtain numerically the distorted wave functions and the differential cross sections. Some that have featured importantly at various times are
   GBROW by Griffy, Biedenharn, Reynolds, Onley and Wright,
   DUELS by Ziegler and the Yale group,
   HEINEL (and FOUBES, son of HEINEL) by Heisenberg,
   HADES by the Mainz group (Andresen, Müller, Peter, Weber).

A typical result (the typical result always illustrated) is that of the excitation of the first excited state of ^{208}Pb, an electric octupole transition, Fig. 5a. The differential cross section, Fig. 5b, has been measured over a range of 10^{11}, and beautifully illustrates the characteristic features: a regular diffraction structure with filled-in minima. The deduced transition charge density, Fig. 5c, is shown together with the uncertainty due to the data points (the thickness of the experimental line!). Also shown are some of the most elaborate theoretical calculations of this quantity. These are not the topic of this talk today, except to say that it is the small lack of agreement with experiment which helps to persuade nuclear theorists that they do not yet understand completely the inside of this nucleus.

We shall hear from Heisenberg of other comparable measurements. At this level of analysis, everything is well under control and there is much useful experimental information of this kind available.

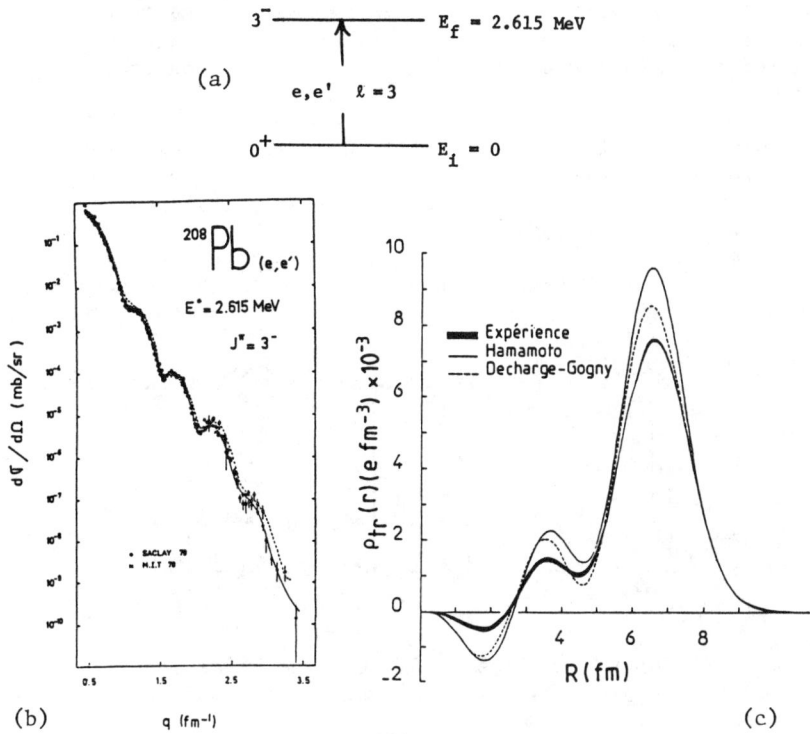

Fig. 5 (a) level scheme for ^{208}Pb transition; (b) differential cross section; (c) transition charge density.

Multiple Exchange of Photons

The Feynman diagram corresponding to the DWBA calculation we have just mentioned is shown in Fig. 6. It contains the exchange of an arbitrary number of 'soft' $\ell=0$ photons associated with the ρ_0 of the nuclear ground state, and correspondingly for the excited state. In the middle, there is the exchange of the 'hard' photon of the requisite multipolarity ℓ to excite the nucleus. Any nucleus will have a variety of nuclear states which can be excited in this manner. It is then fairly clear that it is possible to have contributing diagrams in which several hard photons are exchanged, in the course of which the nucleus is virtually excited to other states. All that matters for the excitation I_i to I_e is that the nucleus starts and finishes in the correct states. Calculations of

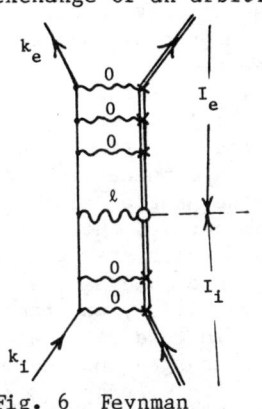

Fig. 6 Feynman diagram for DWBA.

this kind have been reported by Rawitscher, and by Toeppfer and Greiner. Those I use for illustration today are the result of a coupled-channels computer program designed and constructed by Mercer.[3]

Given the wealth of possibilities that are opened up, one looks for the situations in which multiple excitations may contribute to an appreciable degree. The obvious cases are nuclei which are strongly deformed, since then the transition charge densities $\rho_\ell(r)$ which generate the photons are at their largest. The level schemes in Fig. 7a show a chosen deformed nucleus, ^{152}Sm. The diagram in Fig. 7c defines the language

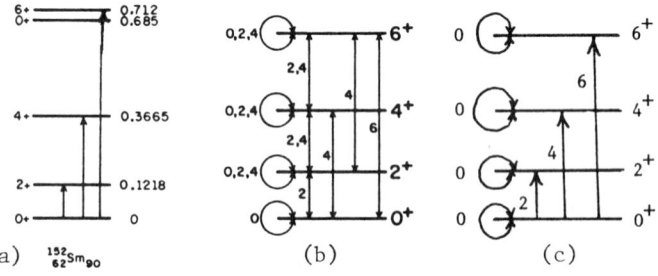

Fig. 7 For the nucleus ^{152}Sm, (a) the level scheme, (b) the coupled-channels coupling scheme, and (c) the DWBA coupling scheme.

used, in that it represents a DWBA calculation of excitation of states in the ground-state rotational band observed experimentally. The numbers label the multipolarities of the transitions, and the soft photon contributions which distort the electron wave functions but which do not change the nuclear state are indicated by the loops labelled 0 (ℓ=0). A deformed-rotor model of this nucleus in which the intrinsic shape is assumed to have ℓ=2, 4 and 6 deformations (so as to give the direct excitation of the states I=2, 4 and 6) contains, subject to angular momentum conservation, matrix elements of the charge operator of these multipolarities between any two of the states. These other couplings are shown in the level scheme in Fig. 7b. (Given the intrinsic shape, these are now prescribed, non-adjustable couplings.) They include ℓ=2, etc moments of the excited states, which can all contribute to the scattering amplitude. Mercer's coupled-channels program permits them all to be included, as interactions which are then calculated to all orders. (It is like a Tamm-Dancoff calculation of a dynamical system: one prescribes the states and their couplings, and then makes an exact 'diagonalization'.) To emphasize this point we show in Fig. 8 some diagrams of hard-photon exchanges implied by this coupling scheme, omitting for simplicity the soft-photon exchanges, which are always with us. (The diagonal quadrupole interaction is also included in non-relativistic Coulomb excitation, where it gives rise to the reorientation effect.)

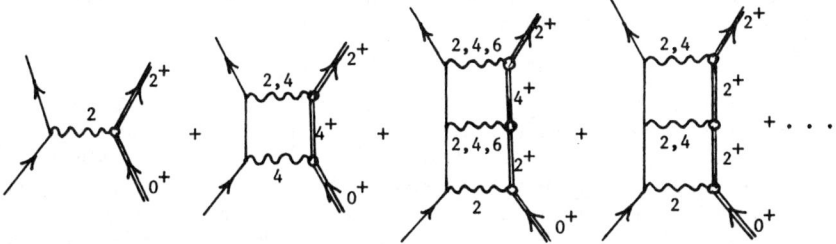

Fig. 8 Some of the Feynman diagrams included when using the coupled-channels method with the coupling scheme of Fig. (7b).

Experimental data on this nucleus taken at Saclay was analyzed by an Illinois group[4] in 1978 using Mercer's coupled-channels program in this manner. In Fig. 9a some sample

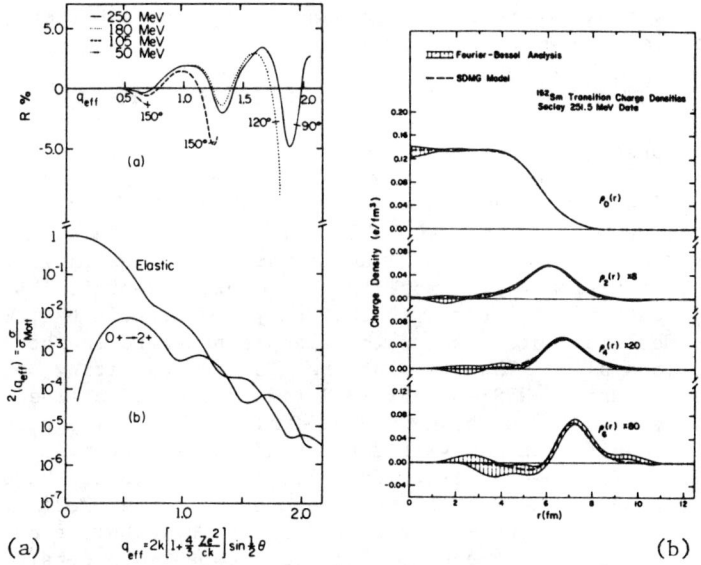

Fig. 9 (a) sizes of coupled-channel effects in ^{152}Sm, (b) deduced transition densities.

calculations show the relative magnitude of the contributions to the cross section extra to those of DWBA. They can become of order 5%. The true measure of their importance in this process, however, is the change they cause in the deduced values of the deformation parameters. These changes turned out to be of the same order as the experimental uncertainties in these parameters (of order 2% for β_2 and 4% for β_4). The resulting charge densities are shown in Fig. 9b. We see that at that stage (which is nearly ten years ago so far as the experiments are concerned), these corrections are good to know about, but they do not modify in an important way the results of a DWBA analysis.

The physical situations covered by Mercer's coupled-channel calculation are those in which the electron-nuclear dynamics is dominated by a few low-lying levels. A computational limitation in that program has been the neglect of the nuclear excitation energy, the effects of which have been estimated to be negligible in the work we have just described. Another source of corrections, examined in the early days by Schiff, is the cumulative effect of virtual excitation of all nuclear states. By using the closure approximation he related this correction to the amplitude to the second Born scattering from the nucleon-nucleon correlation function. More recently, Friar[5] has reported on improvements to this treatment.

It is clear that a complete treatment of the problem needs to combine these approaches. Such a program has not been reported yet, to my knowledge.

EXPERIMENTAL SEARCHES FOR DISPERSION EFFECTS

The presence of virtual-excitation effects, or dispersion effects, may be expected to show up as a small energy-dependent contribution to the basic scattering pattern, which is well described by DWBA. The most likely place to find this contribution is where the DWDA amplitude is smallest, i.e. in a diffraction minimum. The most favorable experiment is ^{12}C elastic scattering[6,7] (I=0, so that only ℓ=0 contributes; low Z, so that Coulomb distortion, which also is non-zero in the minima, is small). The $F(q)^2$ of the elastic scattering, as obtained by Mainz, NBS and now NIKHEF data plotted against a common q_{eff} scale, is shown in Fig. 10a. The mutual agreement among these different sets of data, taken under varying conditions, is shown in Fig. 10b. It is a triumph of the data-analyser's art. The results, at energies up to 240 MeV, are all in essential agreement with the DWBA fit. The NIKHEF cross section at double this energy, Fig. 10c, appears to depart from the DWBA fit in the region of the diffraction minimum by about 8%. Theoretical results of Friar, for this nucleus but somewhat different energies, Fig. 10d, predict a contribution of about 1%. (The apparent spike at the minimum is the result of dividing by the DWBA cross section. The dispersion contribution itself is presumably smooth in that region.) Here, apparently, this admittedly small correction is not yet predictable to within a large factor. It seems to me that the ball is in the theorists' court!

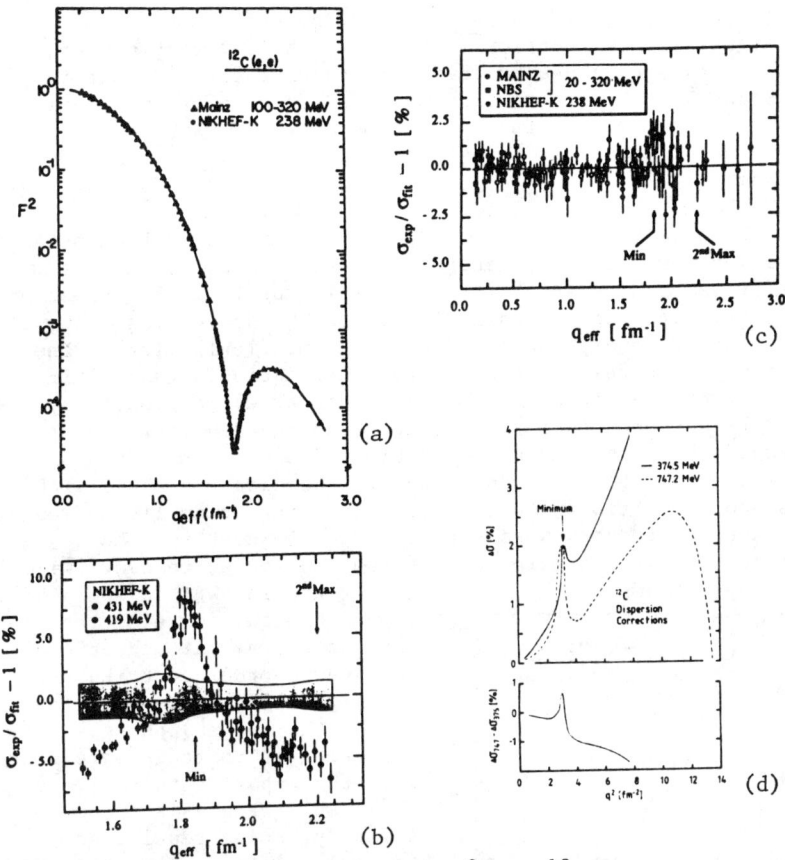

Fig. 10 (a) elastic scattering $(F^c)^2$ for ^{12}C; (b) combined fit of data at or below 240 MeV; (c) comparison of (b), shaded area, with higher energy data; (d) theoretical results reported by Friar[5], at somewhat different energies.

Comparing e^+ Scattering with e^-

Each photon exchange contains as a factor the charge on the incident electron. For the corresponding diagram for positron scattering, there will be +e instead of -e as a factor for each photon exchange. In a phase shift analysis, the effect can be understood most easily by the eikonal calculations we described earlier: the effect of all the soft-photon $\ell=0$ exchange diagrams which constitute elastic scattering is merely to change the sign of the Coulomb potential. The eikonal pictures of Fig. 4 are changed to those corresponding to a diverging lens. The local wave number is <u>reduced</u> by V_C. The effect on the differential cross section is to shift the diffraction structure from its PWBA q-values in the opposite direction to that for electrons. Such effects were observed at Stanford many years ago. As experimental facilities improve, and experimental errors become smaller, it is desirable to look again at the problem.

The ability to fit an electron experimental differential cross section at a given energy with a parametrized charge density using a phase shift analysis is not in itself proof of the sufficiency of this description of the scattering process. It is rather an expression of the extreme flexibility of the fitting procedure. If there were contributions from multi-photon exchanges involving virtual nuclear excitation, they would be subsumed into the effective charge density. The virtual excitation and de-excitation of one state, a two-photon process, would be of second order in the electron charge, and thus would contribute to elastic scattering an amplitude roughly the same for positrons as for electrons (except for the q_{eff} modification). The main part of the amplitude, however, will be different (in PWBA it would be of opposite sign) for electrons and for positrons. If, therefore, the second order contributions are significant, one would not expect the (e^-,e^-) phase shift analysis fit to be in agreement with the experimental (e^+,e^+) phase shift analysis. Some preliminary Saclay results on (e^-,e^-) and (e^+,e^+) scattering from ^{208}Pb at 450 MeV[7] are shown in Fig. 11. It will be interesting to see if the final phase shift analysis fits show any traces of the multi-photon exchanges.

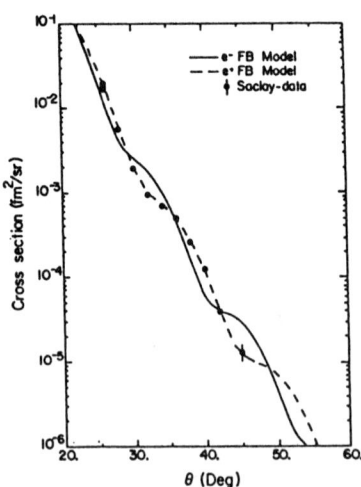

Fig. 11 Fit to (e^-,e^-) on ^{208}Pb, theoretical calculation of corresponding (e^+,e^+), and experimental data.

In inelastic scattering, where the basic DWBA amplitude is smaller, one might expect to see a larger $e^+:e^-$ difference. So inelastic scattering may be a better place to look for this effect.

MORE EXCLUSIVE REACTIONS: ORIENTED TARGETS

One obtains a different view of electron scattering by using an oriented target, especially for the case of a strongly deformed nucleus. This was appreciated and exploited by

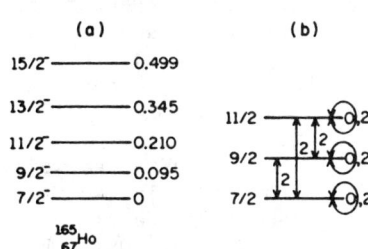

Fig. 12 (a) ^{165}Ho level scheme, (b) coupling scheme for coupled-channels calculation.

Stanford groups a long time ago. The chosen nucleus, ^{165}Ho, has the level scheme shown in Fig. 12a. The inelastic scattering from the members of the ground state rotational band cannot be resolved. The effect of nuclear orientation is to hold the nucleus in a given direction, and with orientation perpendicular to the scattering plane, for example, the unresolved scattering is effectively that from the equatorial cross section of this prolate nucleus. The difference between this and the scattering from the unoriented nucleus is shown in Fig. 13a, together with results obtained for orientation along the $\vec{q}$ direction, which look at the long axis of the football (Fig. 13b).

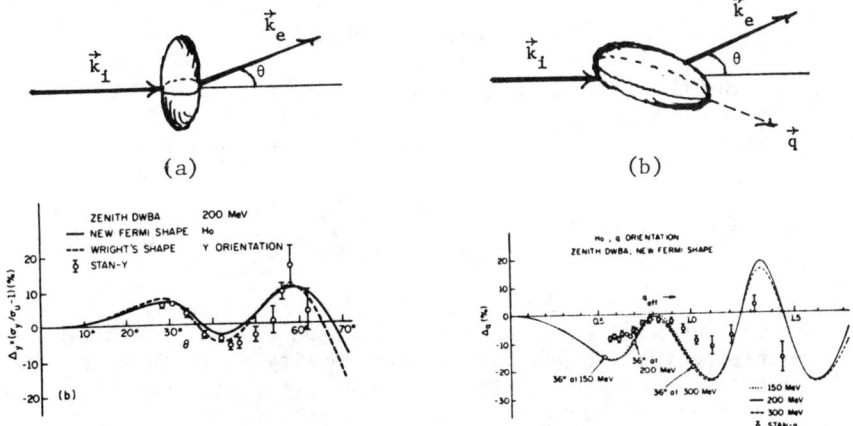

Fig. 13 Geometry and orientation effect for scattering from ^{165}Ho oriented perpendicular to scattering plane (left), and along the $\vec{q}$ direction (right).

In investigating an apparent discrepancy in the analysis of this experiment, using Mercer's coupled-channels program with the simple coupling scheme shown in Fig. 12b, we calculated[8] the expected orientation effect for resolved inelastic scattering. Shown in Fig. 14 are the DWBA and coupled-channels results. One sees that for the 7/2 to 11/2 transition, where there is the possibility of a two-step as well as a one-step excitation, the extra effect of that process is comparable to the DWBA cross section itself. This makes a point we mentioned in the introduction: in this more exclusive process the usually neglected

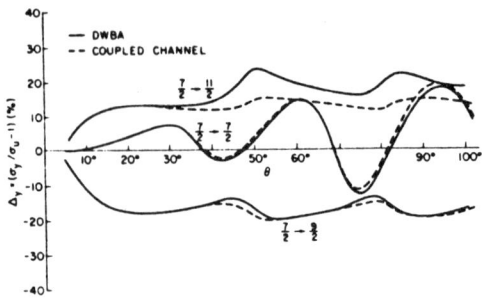

Fig. 14 Predicted orientation effect for resolved elastic and inelastic scattering from ^{165}Ho at 200 MeV.

coupled-channels contributions appear to be relatively larger. As yet, this is all the information we have on this point.

It illustrates well, also, the point we made concerning the e+:e- experiments. While the DWBA curve for the (e+,e+') version of this figure will be the same as those for (e-,e-'), the coupled-channels curves for (e+,e+') would lie on the other side of the full curves from the dashed (e-,e-') curves. With this experiment one would see a big e+:e- effect! Any takers?

NUCLEAR CURRENTS

For the Feynman diagram of Fig. 1, the passing electron represents a current as well as a charge. There is thus an interaction with the nuclear current distribution

$$\langle I_e M_e | \vec{j}(\vec{r}) | I_i M_i \rangle = \sum_{\ell, L=\ell \pm 1} C^{I_i \ell I_e}_{M_i -m M_e} \frac{1}{\sqrt{(2I_e+1)}} j_{\ell L}(r) \vec{Y}^m_{\ell L}(r) \quad (5)$$

as well as with the charge distribution (1). For a given multipole ℓ and transition charge density $\rho_\ell(r)$ there are two transition current densities involved, $j_{\ell\ell-1}(r)$ and $j_{\ell\ell+1}(r)$. They affect the differential cross section through a form factor $F^j_\ell(q)$ in PWBA:

$$F^j_\ell(q) = \sqrt{\frac{\ell+1}{2\ell+1}} F^j_{\ell\ell-1}(q) + \sqrt{\frac{\ell}{2\ell+1}} F^j_{\ell\ell+1}(q), \quad (6a)$$

$$F^j_{\ell\ell\pm 1}(q) = \sqrt{4\pi} \int_0^\infty r^2 dr \, j_{\ell\pm 1}(qr) \, j_{\ell\ell\pm 1}(r). \quad (6b)$$

Part of the current density is just due to the time-variation of $\rho_\ell(r)$, because of charge-current conservation, and thus its contribution to $F^j_\ell(q)$ duplicates the information contained in the charge contribution to the differential cross section. (This is because of the low-q relationship

$$F^j_\ell(q) \to -\frac{\omega}{q} \sqrt{\frac{\ell+1}{\ell}} F^C_\ell(q), \quad q \to 0, \quad (7)$$

which involves only $F^j_{\ell\ell-1}(q)$.) It is $j_{\ell\ell+1}(r)$ which is not constrained by this relationship, and it is the function of q which contains new information about nuclear structure. As with a Rosenbluth analysis of magnetic scattering, the effects of the currents are seen by measuring differential cross sections at the same q-value but at both small and at large angles. It is at the latter that the form-factor F^J_ℓ contributes. This is a complex process involving differences of cross sections, and thus requires the improved accuracy that the present generation of machines provides.

Heisenberg and his collaborators[9], with the computer program FOUBES, make a DWBA analysis of differential cross sections obtained at the Bates accelerator to deduce transition densities $\rho_\ell(r)$ and some $j_{\ell\ell+1}(r)$. Thus at DWBA level this new ingredient in electron-scattering is well under control, at least for (e,e').

MORE COMPLICATED EXCLUSIVE REACTIONS

We have already mentioned the pioneering Stanford work with an oriented target. At some accelerators it is now possible to have a polarized electron beam, which means of course electrons in only one helicity state. By looking at decay products it is also possible, in an inelastic scattering, to examine the state of polarization of the final nuclear state. Let us try to see in physical terms what these new possibilities imply.

We look at the scattering process in the Coulomb gauge. (The gauge used for the electromagnetic field will not affect the results obtained, which must be gauge-invariant. But it can usefully simplify the picture.) The Moller Coulomb potential of the passing electron is $\propto \exp(i\vec{q}\cdot\vec{r})$. Thus with the nuclear quantization axis along the q-direction, the ΔM communicated to the nucleus because of the Coulomb interaction is zero. In this gauge the Moller vector potential is entirely transverse, so that the current interaction communicates $\Delta M = \pm 1$. In Fig. 15 we show M-states for initial and final states of two simple excitation processes, and the way the interactions connect them. Polarizing the spin-1/2 initial nucleus along q has the effect of reducing the M-state complexity to that of a spin-0 nuclear target. In Fig. 16 is shown schematically the effect, at large electron angles, of having electrons in single helicity states. Since helicity is

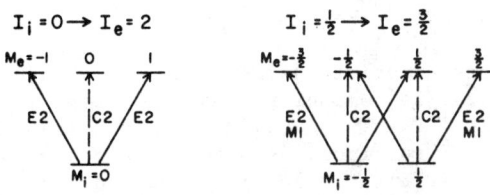

Fig. 15 M-states for sample excitations.

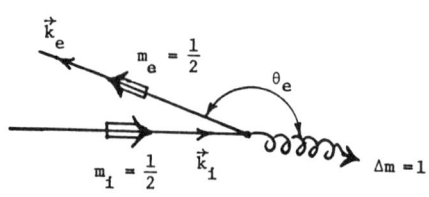

Fig. 16 Schematic diagram of large angle scattering of electrons in the m=1/2 helicity state.

conserved for zero-mass electrons (here an approximation we make for clarity, but not to be repeated in the actual analysis!) only ΔM=1 is communicated to the nucleus, simplifying the M-state populations in a very useful way. A combination of beam and target polarization will then go even further in simplifying the scattering process, allowing thereby the separation of the various competing multipoles.

A non-trivial difficulty with these increasingly exclusive reactions is to get all of the signs and phases in the calculation, PWBA, DWBA or coupled channels, right. (In the simpler (e,e') process the squaring of the amplitude can cover a multitude of signs!)

Final Nuclear M-States from Decay Photons: (e,e'γ)

It is now possible to make coincidence experiments of the type (e,e'x). The angular distribution of the decay product x gives information about how the excitation process populated the excited state, and how that state's structure is related to the fragment x. Decay photons are very convenient to have from the theoretical point of view, since because they do not interact with the rest of the nucleus on their passage out their angular distribution reveals most clearly the M-state population. The spin-density matrix of the excited state has elements which depend linearly on the current form factor F^j (compared to the differential cross section, which depends on $(F^j)^2$). It is therefore possible to detect both the magnitude and the sign of F^j (with respect to that of the charge form factor F^C). This was realized and explored in PWBA by Rose many years ago.

The possibilities of such experiments may be illustrated by looking at the electron scattering community's most popular transition, the excitation in ^{12}C of the 4.44 MeV state. The excitation is induced by the C_2 multipole (charge interaction) and the E_2 multipole (current interaction). The photon decay is purely E_2. The geometry and level scheme are displayed in Fig. 17.

The four-lobed pattern shown in Fig. 18 represents the angular distribution of the decay photons in the scattering plane. In three dimensions the photon pattern is a figure of revolution about the indicated axis. With C_2 excitation alone one would obtain the dashed pattern, which has the $\vec{q}$ direction

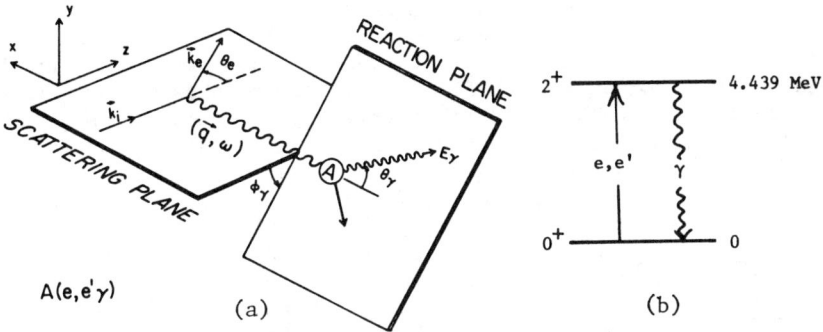

Fig. 17 (a) geometry of the (e,e'γ) process, and (b) nuclear level scheme.

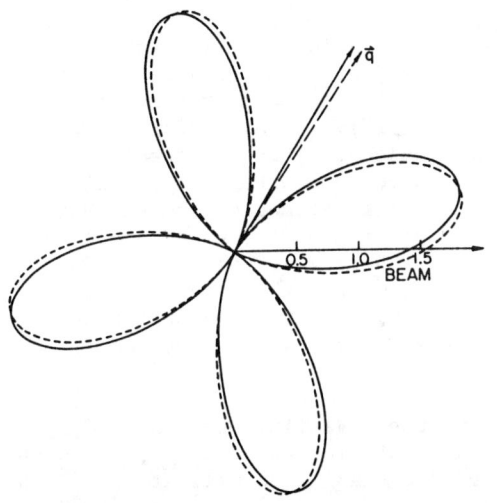

Fig. 18 For ^{12}C 2^+ transition, the photon angular distribution in the scattering plane.

as an axis of symmetry. The full curve is what is actually observed under experimental conditions used in an Illinois experiment[10], with incident electron energy 66.9 MeV, electron angle 60°. The photon pattern is rotated very slightly about 2° from the $\vec{q}$ direction, because of F^J, i.e. because of the interaction of the electron with the nuclear currents. The visibility of this small shift in angle is improved by detecting photons in the regions of the zeroes of the pattern. The Illinois experiment, a demonstration of the possibilities of this technique, showed also that the sign of the current form factor F^J, with respect to that of the charge, F^C, is negative at this low q-value (about 0.3 fm^{-1}), and of a magnitude expected from higher q(e,e') measurements. Direct measurements of $(F^J)^2$ from (e,e') cross sections cannot at present attain such low q-values, of course, because of accuracy limitations.

This technique, in the present application, has measured the phase and magnitude of a current form factor, and it has extended considerably the q-range accessible. This useful

prototype result is obtained by measuring a 2° shift with
necessary accuracy. The question of competing physical effects
which can rotate the pattern is therefore important. A group of
theorists and experimenters at Illinois has been extending these
considerations to DWBA.[11] It is fairly clear from arguments
based on the eikonal approximation that Coulomb distortion of
the electron wave functions can cause a deformation and a
rotation of the photon pattern. We can report that when a DWBA
analysis is made of this reaction, that deformation is a few
percent, and the extra rotation
induced is only of order 5
percent. Shown in Fig. 19 are
the rotations expected, the PWBA
results dashed and those of DWBA
in full, with curves showing how
the phenomenon varies with
electron energy (it decreases
roughly as 1/E) and with nuclear
excitation energy ω (the total
rotation increases as ω, but the
distortion correction not so
rapidly). These results were
made possible by 'stealing'
accurate partial wave radial
matrix elements from the Mainz
program HADES, whose usefulness
we hereby attest to. We do not
know of any way to predict
simply what these corrections
will be, except by calculating
them.

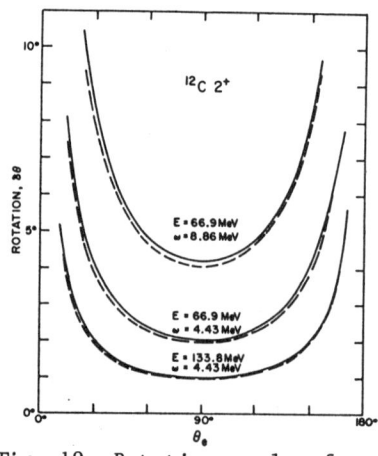

Fig. 19 Rotation angle of
photon pattern for 2^+ ^{12}C
excitation, at different
electron energies and
excitation energies.

We note here that the important quantity observed, F^j, is
the form factor of the total current, including $j_{\ell\ell-1}$, the part
predictable from convection of the charge, as well as $j_{\ell\ell+1}$, the
part one is really interested in.

(e,e'γ) for large Z

In order to see what Coulomb distortion does for transitions at large Z, we can apply our knowledge to the second most
popular (e,e') transition, that to the 2.615 MeV state in ^{208}Pb,
and predict what may be seen in an (e,e'γ) experiment.

For the same conditions as for the ^{12}C experiment, the
photon pattern for this ℓ=3 transition is expected to have the
six-lobed pattern shown in Fig. 20. A comparison of the dashed
curve (C_3 excitation in PWBA) and the full curve (C_3 and E_3
excitation, the former in DWBA) shows that the currents have
produced a rather smaller rotation. This is because of the
smaller excitation energy. (It has been assumed here that the
current is entirely convective. This is compatible with the

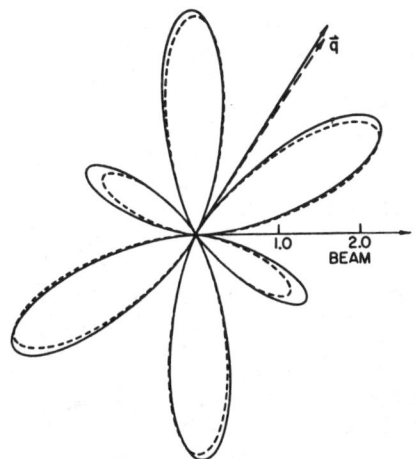

Fig. 20 Photon angular distribution in the scattering plane, for ^{208}Pb 3^- transition, with E_i = 66.9 MeV, θ_e = 60°.

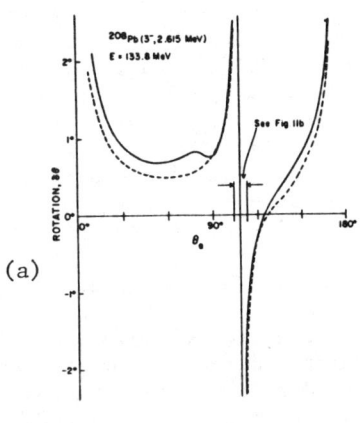

(a)

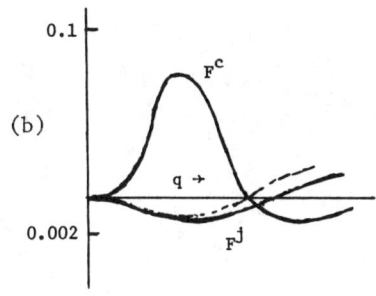

(b)

Fig. 21 (a) expected rotation angle for 3^- ^{208}Pb transition at E_i = 133.8 MeV and (b) sketch of form factors

most recent measurements.) For this large Z, about 30% of this rotation is produced not by the current but by the Coulomg Distortion. The distortion may be considerable: the in-plane to out-of-plane ratio of the photon intensity is about 1.3, under the conditions of Fig. 20, a useful number if one wishes to infer the whole pattern from photon measurement at only a few angles. It seems that here the effects of DWBA over a PWBA calculation are large: this tends to support our suspicion that they are larger the more exclusive the reaction.

Detection of $j_{\ell\ell+1}(r)$

The (e,e'γ) method is not yet fully explored, and while the DWBA corrections are very necessary, there are possibilities for surprises even at the PWBA level. It is easy to show that if the 'interesting' part of the current, $j_{\ell\ell+1}(r)$, is zero, then

$$F_\ell^j(q) = -\frac{\omega}{q}\sqrt{\frac{\ell+1}{\ell}} F_\ell^C(q) . \qquad (8)$$

Consequently at a diffraction minimum both F^C and the convective contribution to F^j are zero. One may search there for effects of even a small $j_{\ell\ell+1}$. In Fig. 21a is shown the rotation angle calculated for the 3^- transition in ^{208}Pb, at an energy of 133.8 MeV, double that of the previous discussion. The first diffraction zero is at $q=1.1$ fm^{-1} or $\theta=108°$. To simulate an unknown $j_{\ell\ell+1}(r)$ we have assumed that $F^j(q)$ is that of pure convection, but with the scale of q changed by 10%. (It is equivalent to assuming a form for the current which has the same structure as the convection part, but with a smaller radius.) The form factors are sketched in Fig. 21b. It is evident from the plot of the rotation angle that something drastic happens at the diffraction zero: the rotation angle appears to go off scale. What is actually occurring is shown in Fig. 22. In the three degrees of electron scattering angle spanning the diffraction minimum, the six-lobed decay photon angular distribution has rotated through one whole lobe. (Despite appearances, the patterns shown in this progression are each normalized to 4π over all solid angles of emission.)

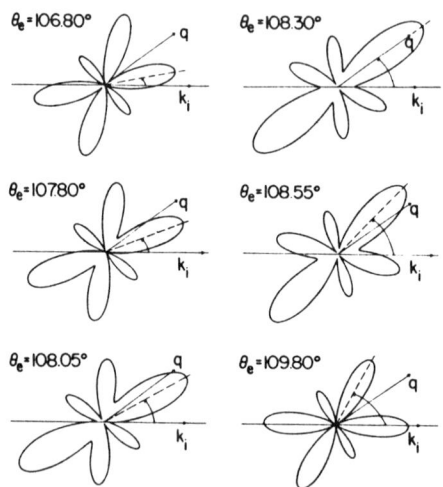

Fig. 22 In-plane photon angular distributions for the 3^- ^{208}Pb transition at electron angles spanning the diffraction minimum.

The way this has happened is as follows: at most electron angles F^C dominates, and the ρ_{00} density matrix element is close to 1. This has the effect of anchoring the photon pattern at the normal one of Fig. 19a, with axis in the $\vec{q}$ direction. At the minimum, F^C becomes the small effect, and the currents alone make the pattern shown at 108.05° electron angle, which is a different D-function and has a different shape - although still a six-lobed pattern in the scattering plane, with q-direction an approximate axis of symmetry. One hopes that this dramatic change in photon pattern will be detectable. It will be an indicator of the presence of $j_{\ell\ell+1}$, which is hard to detect otherwise.

We note that while our discussion has been couched in PWBA language, the results shown are in DWBA. Otherwise one would not be sure of a discussion concerning a diffraction minimum, a place where PWBA is most deficient. As it is, one would like to see a coupled-channels treatment, so as to be sure that there are not some other corrections.

AN INTERMEDIATE ENERGY EXAMPLE: STRUCTURE OF THE n-Δ TRANSITION

To show that while our accelerator energy is limited, our ambitions are not, we mention without results an obviously interesting topic at a higher excitation energy. This is the excitation of the Δ-resonance of the nucleon by electrons, and the possible presence of multipole components which would bear on the question of the quadrupole deformation of that state. The two diagrams which are central to the discussion are shown in Fig. 23. The M-states of the proton and the Δ are those shown in Fig. 14b. At question is the size of the C_2 excitation amplitude. It is interesting because one would like to have an experimental value, or even an upper limit, for the quadrupole moment of the Δ. Popular models for the nucleon ground and excited states, such as the Bag Model in its various forms, or the Skyrmion, show a tendency towards quadrupole deformation.

Unfortunately almost all of our techniques and insights need drastic improvement in order to treat this problem. One cannot separate the excitation from the decay, as there are crossed graphs contributing. There are several other background diagrams which are experimentally indistinguishable from the ones of interest. Etc. (At least, Coulomb distortion effects will be small!)

Nonetheless, one would think that it is possible by careful measurement of the decay photons and those from the neutral pion to detect the multipoles of interest. From what we have learned from nuclei, it seems clear that the process would be very much simplified by the use of polarized electrons and a polarized proton target. We do not wish, however, to minimize the theoretical difficulties lying in the path from detection of C_2 to a deduction of a quadrupole moment.

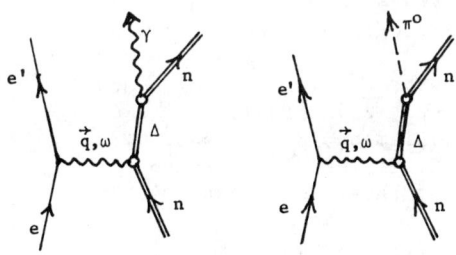

Fig. 23 Diagrams for examining the n-Δ transition.

SUMMARY

We have tried to discuss the increasingly exclusive, increasingly precise kinds of experiments that electron scattering is now focussing on, and the necessity for calculations of sufficient scope and accuracy in each case. It appears from the fragmentary evidence cited that the corrections called for by Coulomb distortion and nuclear dynamics may be larger as the processes become more exclusive. It is only by carrying through such calculations that we can all be confident of the precision of the electron probe.

ACKNOWLEDGEMENTS

It is a pleasure to acknowledge discussions with C. N. Papanicolas and L. S. Cardman during the hurried preparation of this talk. Research by D. G. R. was supported in part by the National Science Foundation under grant NSF PHY84-15064.

REFERENCES

1. D. R. Yennie, F. L. Boos, Jr. and D. G. Ravenhall, Phys. Rev. 137B, 882 (1965).
2. D. Goutte, J. B. Bellicard, J. M. Cavedon, B. Frois, M. Huet, P. LeConte, Phan Xuan Ho, S. Platchkov, J. Heisenberg, J. Lichtenstadt, C. N. Papanicolas and I. Sick, Phys. Rev. Lett. 45 1618 (1980), and private communication.
3. R. L. Mercer, Phys. Rev. C15, 1786 (1977).
4. L. S. Cardman, D. H. Dowell, R. L. Gulbranson, D. G. Ravenhall and R. L. Mercer, Phys. Rev. C18, 1388 (1978).
5. J. L. Friar, in Electron and Pion Interactions with Nuclei at Intermediate Energies, ed. W. Bertozzi, S. Costa and C. Schaerf (Harwood, New York, 1980).
6. E. A. J. M. Offerman, L. S. Cardman, H. J. Emrich, G. Fricke, C. W. de Jager, H. Miska, D. Rychel and H. De Vries, Phys. Rev. Lett. 57, 1546 (1986).
7. B. Frois and L. Cardman, private communication.
8. D. G. Ravenhall and R. L. Mercer, Phys. Rev. C13, 2324 (1976).
9. J. Heisenberg, J. Lichtenstadt, C. N. Papanicolas and J. S. McCarthy, Phys. Rev. C25, 2292 (1982).
10. C. N. Papanicolas, S. E. Williamson, H. Rothaas, G. O. Bolme, L. J. Koester, Jr., B. Miller, R. Miskinen, P. Mueller and L. S. Cardman, Phys. Rev. Lett. 54, 26 (1985).
11. D. G. Ravenhall, R. L. Schult, J. Wambach, C. N. Papanicolas and S. E. Williamson, Annals of Physics (in press).

SESSION D

Higher Resolution and Higher Momentum Transfer: New Insight into Nuclear Structure

by

J.H. Heisenberg

Dept. of Physics, University of New Hampshire, Durham, N.H. 03824

Introduction

We are here to celebrate the past together with the development that has taken place during the last thirty years. In this period I have personally participated only during the last twenty years after I came to Stanford University as a young research associate. At that time electron scattering was already a celebrated but somewhat isolated field. Yet, since then a tremendous development has taken place that has made electron scattering one of the major forces for advancing and stimulating our understanding of nuclear structure.

Looking back at the experiments done at Stanford, we find that a large portion of the nuclear structure experiments that we consider as significant milestones were already attempted or done at Stanford with the limiting technology available then. The improvements since that time are not only in hardware but also in data analysis techniques that allow to relate the measured cross sections to nuclear structure properties. Since we are concerned here with the development that brought us to the presence I would like to take the first part of my talk to review this development. In the second part I will try to present how electron scattering experiments have affected our understanding of the nucleus.

Establishing the Mean Field Approach

During my two years at Stanford, electron scattering experiments could be done typically with $2\mu A$ of average beam current and resolution of $\frac{\Delta p}{p} = 2 \times 10^{-3}$. For typical energies between 300 and 500 MeV this meant resolution of approximately 1 MeV. For elastic scattering the formalism for the DWBA treatment of the scattering even at high energies had been developed by Yennie, Booth and Ravenhall[1] allowing for a reliable calculation of cross sections for a given charge density. The just completed experiment of Frosch et al.[2] had been analyzed showing that the ground state charge densities of the ^{40,48}Ca nuclei could not be adequately described by a 3-parameter Fermi distribution. This was the first observed discrepancy from a shape that seemed already quite sophisticated.

In inelastic scattering experiments that were carried out mostly at Darmstadt and Yale at low energies for reasons of resolution, the aim was to reliably extrapolate in momentum transfer to the photon point in order to determine the B(Eλ) or B(Mλ) for the transition. In these extrapolations transition radii were determined that were the equivalent to the ground state rms-radii. However, I do not know of a single study where the value of this transition radius was used to clarify some nuclear structure. This demonstrates that the transition probabilities were essentially the only overlap between the conventional nuclear structure and electron scattering results. Also for inelastic scattering the foundations for a DWBA treatment of the scattering had been completed by Griffy, Biedenharn, Reynolds, Onley and Wright[3] in the "GBROW" program. At that time, running such a lengthy program was an expensive proposition and for that reason rare.

The resolution of 1 MeV allowed for some, though limited, nuclear structure program at Stanford, and a considerable number of transitions to excited states were mapped out in momentum transfer up to typically 2.5 fm^{-1}. In most of these studies it was important to find a reasonable model for the transition densities in order to accurately extrapolate to the photon point. At that time we found in the excitation of the collective octupole vibration in ^{208}Pb[4] for the first time that also for inelastic transitions the model of a simple surface peaked shape could not represent the transition density. What was almost more important was that the structures seen in this transition density had a one-to-one correspondence to structures predicted by one of the first microscopic calculations of this transition density by Gillet[5]. This demonstrated that there was a lot of useful information coming out of these measurements that went far beyond the mere determination of the transition probability.

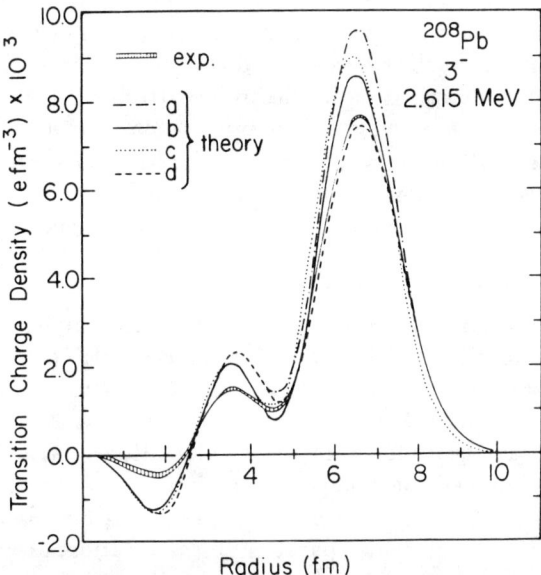

Fig.1 Transition charge density of the octupole vibration in ^{208}Pb[6].

These findings changed the direction of our attention away from the measurement of transition probabilities to the measurement of the full radial structure of the densities. In the next few years, while we were waiting for the Bates accelerator to come on the air, significant advances were made in reconstructing these densities from the data together with errors. These techniques, called "model independent" analyses, were employed to nuclear ground states in form of the Fourier-Bessel expansion by Friar and Negele[7] and in form of the sum of Gaussians by Sick[8]. The Fourier-Bessel expansion technique for reconstructing transition densities was also applied in inelastic scattering for the first time by Rothhaas et al.[9] confirming the shape of the transition density for the octupole vibration in ^{208}Pb as determined earlier. However, an essential element for a meaningful comparison to theory had been added in procedure constructing uncertainties in these densities. This allowed one to get a sense of the significance of some observed discrepancy. Fig.1 shows the results from the most recent analysis of the same octupole transition.

In the following years significant progress had been made in the instruments that became available. The Saclay accelerator was able to produce beams up to 700 MeV with beam currents more than a factor of 10 above typical currents at Stanford. The resolution has been improved and resolution of 1×10^{-4} have been achieved by now. In 1975 experiments started at the Bates accelerator where resolution of 4×10^{-5} have been achieved so far. The energy available is now 800 MeV with typical beam currents up to $60\mu A$. The accelerator at NIKHEF provides beams of 500 MeV and $40\mu A$ and resolutions of 5×10^{-5}. With these facilities nuclear structure programs are no longer limited to a few well separated low lying nuclear states, but the whole spectrum of nuclear levels up to rather large level densities has become available for investigation by electron scattering.

With these machines a wealth of new information came about. The Saclay group extended the very precise cross section measurements from the Mainz accelerator to higher momentum transfers of $\geq 3.5 fm^{-1}$ such that extending the measurements even further would no longer improve our knowledge of the charge density. Fig.2 shows several densities derived that way[10].

Also in the early seventies the development in computer technology allowed for more and more sophisticated Hartree-Fock(HF) calculations that warranted such scrutiny in comparison with experiment. Also shown in Fig.2 are densities calculated with the Gogny D1 force. It is impressive how well nuclear radii are reproduced in these calculations. Also, the surface properties are reproduced extremely well.

At the Bates accelerator a series of experiments on deformed nuclei were done where the insufficient separation of the rotational levels from the Stanford experiment[11] or the restriction to low momentum transfer of the NBS-experiments[12] were no longer a limitation. From these experiments

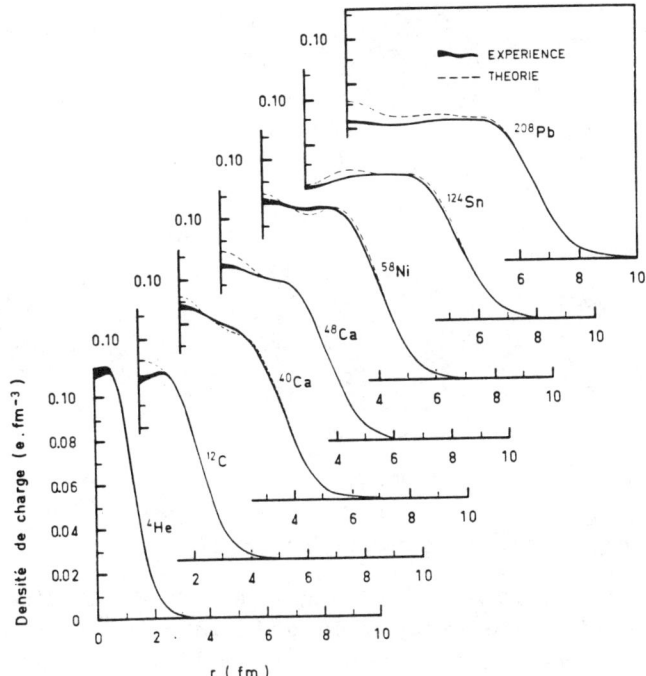

Fig.2 Ground state charge densities compared to HF results.

one could reconstruct the intrinsic deformed charge density of the nucleus. Such a density derived from the experiment of Hersman et al.[13] is shown in Fig.3 for the nucleus ^{154}Gd in form of a contour plot. In comparing these results with HF-calculations one finds again that the surface properties are reproduced rather well, including the deformation of the surface, but that the density in the interior is more smooth than what is predicted by these calculations.

It was the availability of high momentum transfer data that changed comparing measured and calculated rms-radii into comparing the whole shape and in particular the structures in the densities where one finds significant discrepancies, pointing out significant deficiencies in the theory or calculation.

In inelastic scattering the attention moved away from the study of the strong collective states to the study of weakly excited levels where through separation of the longitudinal and transverse form factors it was possible to measure two densities for electric excitations. Together with the measurement of high multipolarity magnetic excitations these results proved to be extremely useful.

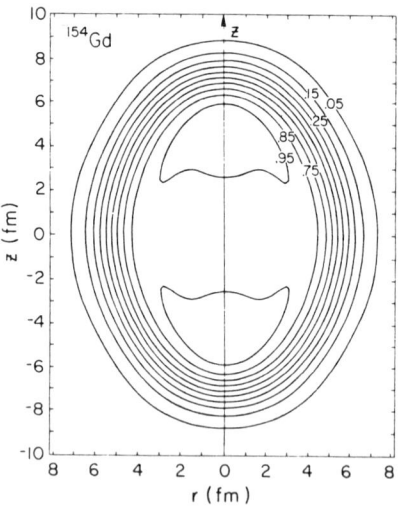

Fig.3 Density-contourplot for ^{154}Gd.

Of particular interest appeared to be the high spin stretched ph-states of which the first ones in ^{12}C were already observed at Stanford[14]. With the new high resolution one was able to resolve most of these levels and to detect even fairly small fragments. In Fig.4 I show the results of the form factor for the highest multiplicity ph-configurations observed so far, the 12^-, and 14^- states in ^{208}Pb[15]. These states are expected to have the simple configurations of $\nu(1j_{\frac{15}{2}}, 1i_{\frac{13}{2}}^{-1})_{14}$ for the 14^- state and $\pi(1i_{\frac{13}{2}}, 1h_{\frac{11}{2}}^{-1})_{12}$ and $\nu(1j_{\frac{15}{2}}, 1i_{\frac{13}{2}}^{-1})_{12}$ for the 12^- states. Even though these seemed to be rather pure configurations, there was a common feature to all of these states, that the strength observed was usually less than half of the strength predicted by such a simple ph-picture. The most simple way to accomodate this feature was to assume that the effective magnetic moment of the nucleon had changed. Thus, since we are no longer dealing with free nucleons but nucleons embedded in a nuclear medium, the properties of the nucleon are modified.

Challenging the Mean Field Approach

Before one takes as radical measures as to blame the quark-structure of the nucleon for the change of the properties one has to study the effects of the nuclear medium itself, and several experiments were designed to shed some light on the dominant effects causing such a reduction in strength.

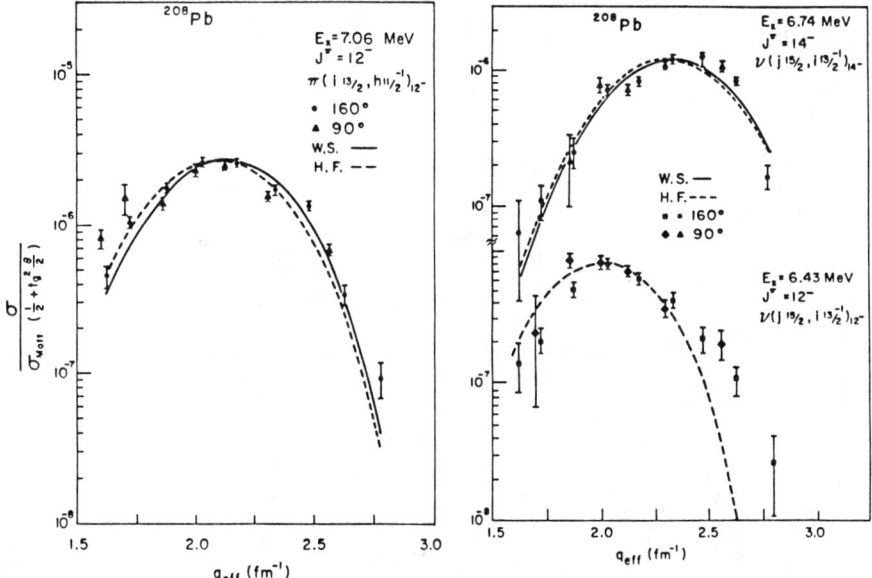

Fig.4 Form factor for the excitation of 14^- and 12^- states in ^{208}Pb

For the low lying states pairing was shown to be a significant factor in this reduction. Let me give you an example looking at E5 excitations in the N=50 nuclei ^{89}Y, ^{90}Zr, and ^{92}Mo. All three nuclei have a low lying excitation of the proton transition $2p_{\frac{1}{2}} \to 1g_{\frac{9}{2}}$. In the HF-picture we would expect an increase in strength in this transition for ^{90}Zr over ^{89}Y as there are twice as many protons available in the $2p_{\frac{1}{2}}$-orbit. As we go to ^{92}Mo we expect a decrease in strength as partial blocking in the $1g_{\frac{9}{2}}$ orbit takes effect.

Some preliminary results for the transition charge densities and current densities for ^{89}Y[16], ^{90}Zr[17], and ^{92}Mo[18] are shown in Fig.5. Indeed, the transition charge densities have the characteristic node expected for the $2p \to 1g$ transition which should be the product of the two radial functions. Both the transition charge densities and the transition current densities look remarkably similar. However, the strength observed does not follow the simple HF picture. The charge density in ^{92}Mo is even larger than that from ^{90}Zr. What is even more serious is, that the strength of the transverse current does not follow the strength of the charge. The only way to explain these strength ratios is by assuming partial occupations. In fact, if pairing

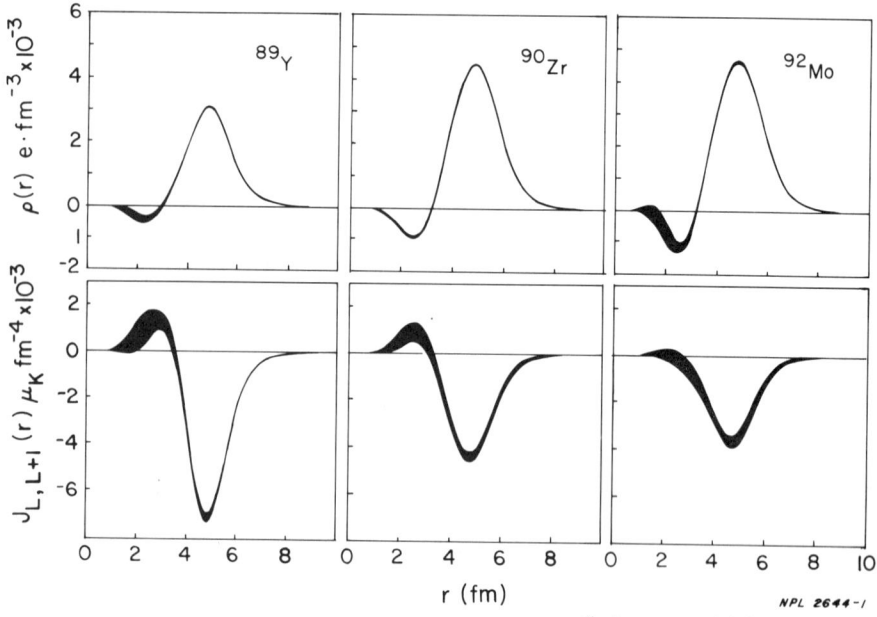

Fig.5 Transition charge (top) and current densities (bottom) for the first E5 transitions in ^{89}Y, ^{90}Zr, and ^{92}Mo

is the major cause for this partial occupancy we have to make a quasi-particle transformation. The single quasi-particle charge and current densities $\tilde{\rho}_\lambda^{ph}$ and $\tilde{J}_{\lambda,\lambda'}^{ph}$ are given in terms of the single particle densities as

$$\tilde{\rho}_\lambda^{ph}(r) = (u_p v_h + v_p u_h)\rho_\lambda^{ph}(r)$$

$$\tilde{J}_{\lambda,\lambda'}^{ph}(r) = (u_p v_h - v_p u_h)J_{\lambda,\lambda'}^{ph}(r)$$

Here $v^2 = 1 - u^2$ is the occupation probability of the respective orbit in the nuclear ground state. From these preliminary results we find a consistent set of occupations n

$$n_{Zr}(2p_{\frac{1}{2}}) = 0.63,$$

$$n_{Zr}(1g_{\frac{9}{2}}) = 0.16,$$

$$n_{Mo}(2p_{\frac{1}{2}}) = 0.74,$$

and

$$n_{Mo}(1g_{\frac{9}{2}}) = 0.33.$$

These numbers indicate a significant softening of the Fermi surface.

A second effect modifying the density-operators is core polarization. This shows up most strongly in the transition charge densities. Core polarization can be understood as a coupling between the single particles and the collective core excitations. The fact that we do not observe collective magnetic transitions indicates already that the direct effects on the currents are small. Core polarization can be reasonably well described already in a first order calculation. Fig.6 shows experimental results and calculations for the multiplet 2^+, 4^+, 6^+, and 8^+ in ^{90}Zr[19] that comes about from the recoupling of a pair of protons in the $1g_{\frac{9}{2}}$ orbit. The calculations are shown without core polarization (dotted line) and with core polarization (full line). This calculation implies that the effects of core polarization decrease rapidly with increasing multipolarity.

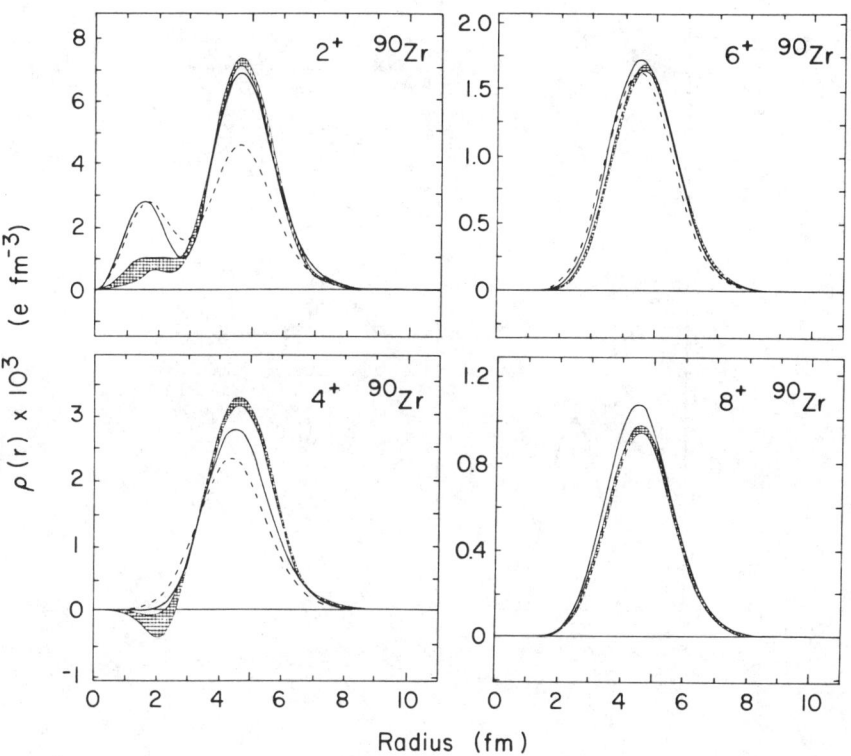

Fig.6 Experimental and calculated transition densities for the 2^+, 4^+, 6^+, and 8^+ multiplet in ^{90}Zr

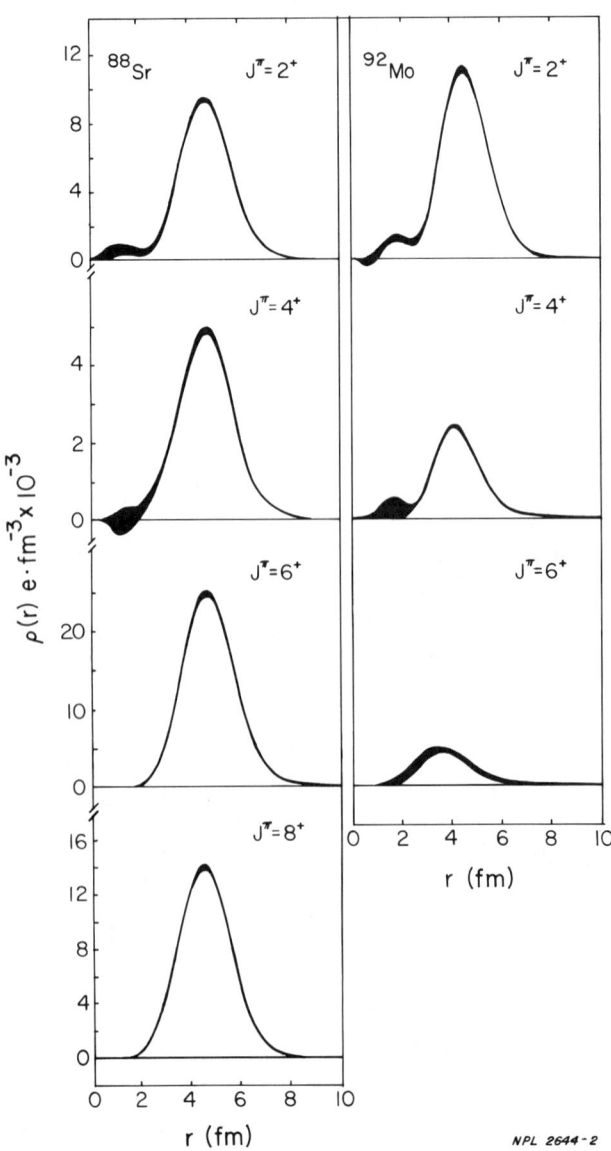

Fig.7 Comparison of transition charge densities of the 2^+, 4^+, 6^+, and 8^+ multiplets in ^{86}Sr and ^{92}Mo.

To verify this effect in a more quantitative way, we can compare such multiplets for proton and neutron configurations. The multiplets most equivalent in structure are from ^{92}Mo with the configuration $(\pi 1g_{\frac{9}{2}})^2$ and ^{86}Sr with the configuration $(\nu 1g_{\frac{9}{2}})^{-2}$. Fig.7 shows preliminary results of the equivalent densities for these two cases. We find that for ^{86}Sr[20] the densities become rapidly smaller than their counterpart in ^{92}Mo[18], and for the 8$^+$ level only an upper limit on the density is obtained which is a factor of ten smaller than the equivalent density in ^{92}Mo.

There is, however, another very significant effect arising from the coupling of the single particle orbits to the collective modes which is again a softening of the Fermi surface. We can study these effects on the single particle transitions in ^{89}Y. Fig.8 shows the transitions $\pi(2p_{\frac{3}{2}} \to 2p_{\frac{1}{2}})$ and $\pi(1f_{\frac{5}{2}} \to 2p_{\frac{1}{2}})$. The single particle predictions shown as long dashed lines do not predict enough strength in the charge density but too much strength in the current density.

The coupling to the collective modes here has two effects: The softening of the Fermi surface brings the spectroscopic strength of this transition down. This is seen in the strong reduction of strength in the charge density in the interior of the nucleus as well as in the reduction of strength in the

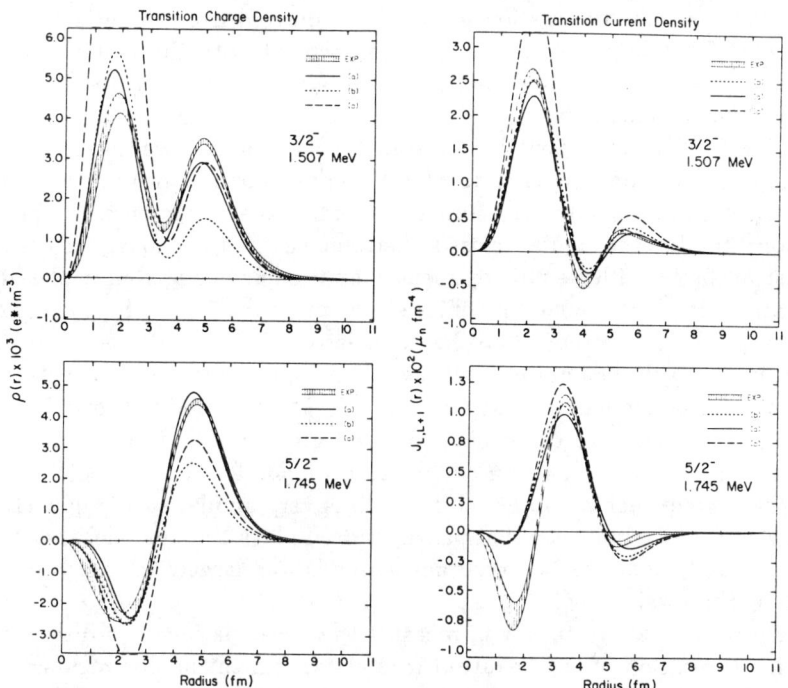

Fig.8 Charge and current transition densities for the 3/2$^-$ (1.507 MeV) and the 5/2$^-$ (1.745 MeV) levels in ^{89}Y.

current density. Core polarization as seen in the previous examples is a dominant surface effect. It compensates the reduction of the spectroscopic strength by adding strength at the nuclear surface such as to generate an overall increase in the transition probability. The effects of core polarization on the currents are very small. With respect to the reduction of the spectroscopic strength, calculations show that about half of it comes from the coupling to the positive parity phonons and half comes from the coupling to negative parity phonons.

One should note that the strength in the transition current is quite well reproduced, indicating that the quenching in the currents can essentially be explained by this softening of the Fermi surface. Thus, a picture emerges where there is substantial depopulation of the occupied HF orbits and substantial population of the unoccupied HF orbits. Electron scattering results can give information allowing to check this picture in a quantitative way. In fact, the quenching of the transition currents observed at high momentum transfer is to the largest part due to this repopulation and thus gives an almost quantitative image of this repopulation.

A complementary way of looking at these repopulations is through knock-out reactions. After the pioneering work at Saclay in this field of $(e, e'p)$ reactions, NIKHEF has more recently produced a lot of rather impressive work. The occupation probabilities derived from these $(e, e'p)$ experiments for ^{90}Zr[21] is in reasonable agreement with the results quoted above. Thus we see some consistency emerging between results from two rather different reactions.

Since ^{208}Pb is somehow the key example of a doubly closed shell nucleus, let me use the remaining time to review the experiments in the Pb-region. The situation is actually very similar to that in the Zr-region with the exception that for the doubly closed shell nucleus ^{208}Pb pairing effects are almost negligible. I have already mentioned the quenching seen in the high spin states in ^{208}Pb, which includes states of $J^\pi = 7^-, 8^-, 10^-, 10^+, 12^-$, and 14^-. The measurements of Papanicolas et al.[22] on the neutron hole states in ^{207}Pb showed quite significant quenching in the transverse scattering form factor similar to that seen in ^{89}Y. The longitudinal form factors are similar to those of the collective core excitations, indicating that the picture of the particle-phonon coupling also holds for the Pb-region. The fact that the quenching observed in ^{207}Pb is very similar to the quenching observed in ^{208}Pb indicates that pairing indeed plays a minor role, and these findings imply that the observed quenching is due largely to the softening of the Fermi level.

There is no accurate way to measure these occupations in a direct way. The interpretation of the beautiful $(e, e'p)$ experiment in this region[23] in terms of absolute spectroscopic information is plagued by some of the same problems encountered in the $(^3He, d)$ reaction, namely the need for understanding the proton reabsorption in a quantitative way. Furthermore, we do not yet have programs available that treat both the distortion in the electron wave and the distortion in the proton wave. However, compared to the $(^3He, d)$ reaction, one is no longer sensitive to only the tails of the wave functions at large distances which certainly eliminates some of the problems. Without this direct determination one is limited to relative measurements increasing the uncertainties substantially due to error propagation. Thus one has to tie the many pieces of the puzzle together to form a complete and consistent picture.

There are many pieces to this puzzle. In addition to the excitation of high spin levels mentioned above, there is the beautiful experiment of the charge density differences between ^{206}Pb and ^{205}Tl of the Saclay group[24]. The charge density difference measurement shown in Fig.9 allows the extraction of the difference in the $3s_{\frac{1}{2}}$ strength due to the very characteristic form factor shape of the $3s_{\frac{1}{2}}$ orbit.

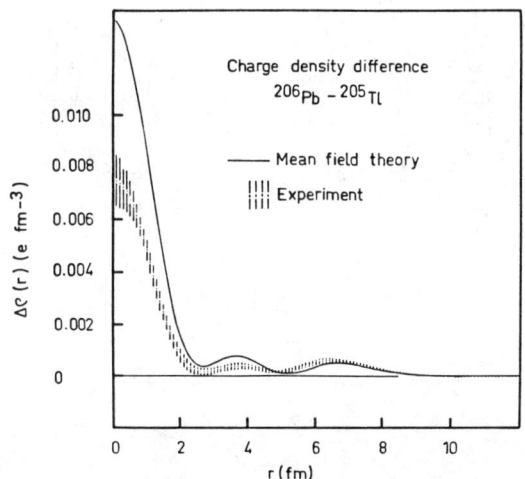

Fig.9 Charge density difference ^{206}Pb and ^{205}Tl.

Again, relating that determination of absolute occupation differences to an absolute occupation probability is complicated by the fact that the nucleus ^{205}Tl can not be considered to be a pure configuration $3s_{\frac{1}{2}}^{-1}|^{206}Pb_{0+}\rangle$. Rather, experiments of ^{206}Pb$(^3He, d)$ or ^{206}Pb$(e, e'p)$ show configuration mixing in the ground state. Calculations of Zamick, Klemt, and Speth[25] show that the wave function of the ground state is more closely represented by

$$|^{205}Tl\rangle = \alpha 3s_{\frac{1}{2}}^{-1}|^{206}Pb_{0+}\rangle + \beta 2d_{\frac{3}{2}}^{-1}|^{206}Pb_{2+}\rangle + \gamma 2d_{\frac{5}{2}}^{-1}|^{206}Pb_{2+}\rangle$$

in agreement with the particle-phonon coupling picture. The uncertainty in the value of α again limits the achievable accuracy.

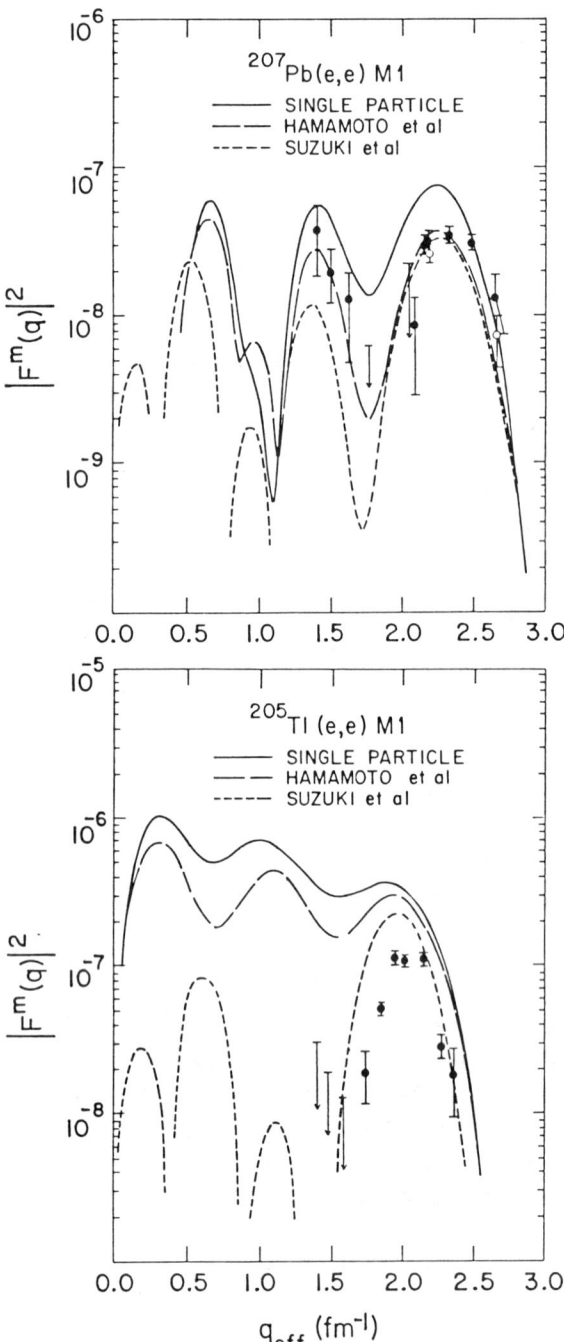

Fig.10 Ground state magnetic form factors for ^{207}Pb (top) and ^{205}Tl (bottom)

These results are complemented by the measurements of the ground state magnetic form factors of ^{205}Tl and ^{207}Pb[26] which do show effects from core polarization. These form factors are shown in Fig.10.

This effect of softening of the Fermi surface in ^{208}Pb was calculated by Pandharipande, Papanicolas, and Wambach[27]. The results are shown in Fig. 11. The authors include three different effects: Short range correlations, tensor correlations, and long range correlations. The experiments discussed so far involve only orbits very close to the Fermi level. In that region, the effects are dominated by the softening due to long range correlations and tensor correlations which are adequately described in the particle-phonon coupling picture. They will not permit us to learn about the short range correlations. For effects from those one will have to look at much higher excitation energies such as the quasielastic peak.

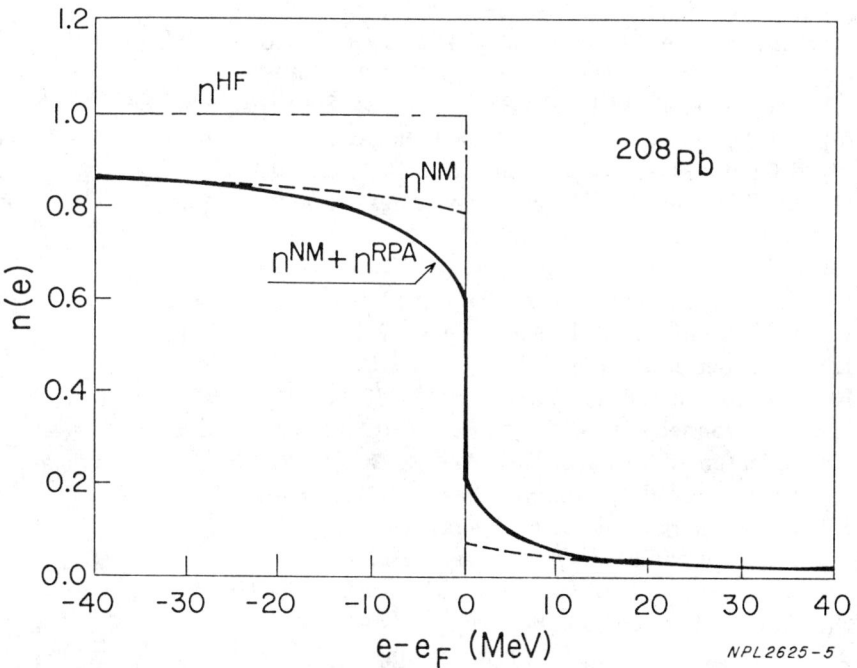

Fig.11 Calculated occupation probabilities in ^{208}Pb.

The extracted occupation probability for the $3s_{\frac{1}{2}}$-orbit in ^{208}Pb with its still large uncertainty is consistent with the estimates of Pandharipande et al.. With all these pieces we are waiting for a unifying calculation that treats all the observed effects in a self consistent approach. Only such a calculation would give credence to the predicted occupations if with it all observables are reproduced in a quantitative way.

In summary, we have seen that the measurements of high momentum transfer data allows to extract properties of very specific orbits due to the characteristic momentum transfer dependence of their respective form factors. To fully use this sensitivity, form factors need to be mapped out up to $\geq 3.5 fm^{-1}$. The bulk of the new results comes from weaker levels where resolution is essential not only to separate the levels from neighboring states but also to enhance the peak to background ratio. Progress in both of these criteria have produced experimental results that force us to go beyond the mean field HF-approach and a picture of the nucleus emerges in which we see substantial repopulations for levels around the Fermi surface even for doubly magic nuclei.

References

(1) D.R.Yennie et al., *Phys. Rev.* **95** (1954), 500.
(2) R.F.Frosch et al., *Phys. Rev.* **174** (1968), 1380.
(3) S.T.Tuan et al., *Nucl. Instr. Meth.* **60** (1968),70.
(4) J.Heisenberg and I.Sick, *Phys. Lett.* **32 B** (1970), 249.
(5) V.Gillet et al., *Nucl. Phys.* **88** (1966), 321.
(6) D.Goutte et al., *Phys. Rev. Lett.* **45** (1980), 1618
(7) J.L.Friar and J.W.Negele, *Nucl. Phys.* **A 212** (1973), 93.
(8) I.Sick, *Nucl. Phys.* **A 218** (1974), 509.
(9) H.Rothhaas et al., *Phys. Lett.* **51 B** (1974), 23.
(10) J.M.Cavedon and B.Frois, Rapport DPh-N Saclay, no 2137
(11) J.H.Heisenberg et al., *Nucl. Phys.* **A 164** (1971), 340.
(12) T.Cooper et al., *Phys. Rev.* **C 13** (1976), 1083.
(13) F.W.Hersman et al., *Phys. Rev.* **C 33** (1986), 1905.
(14) T.W.Donnelly et al., *Phys. Rev. Lett.* **21** (1968), 1196.
(15) J.Lichtenstadt et al., *Phys. Rev.* **C 20** (1979), 497.
(16) J.E.Wise et al., *preliminary results*
(17) J.Heisenberg et al., *to be published*
(18) T.E.Milliman et al., *preliminary results*
(19) J.H.Heisenberg, *Comm. Nucl. Part. Phys.* **13** (1984), 267.
(20) J.Connelly et al., *preliminary results.*
(21) P.K.A.de Witt Huberts, *priv. comm.*
(22) C.N.Papanicolas et al., *Phys. Rev. Lett.* **45** (1980), 106.
(23) E.N.M.Quint et al., *Phys. Rev. Lett.* **57** (1986), 186.
(24) B.Frois et al., *Nucl. Phys.* **A 396** (1983), 409c.
(25) L.Zamick et al., *Nucl. Phys.* **A 245** (1975), 365.
(26) C.N.Papanicolas et al., *Phys. Rev. Lett.* **58** (1987), 2296.
(27) V.R.Pandharipande et al., *Phys. Rev. Lett.* **53** (1984), 1133.

INTERMEDIATE ENERGY NUCLEAR PHYSICS WITH ELECTRONS*

E. J. Moniz
Bates Linear Accelerator Center
Laboratory for Nuclear Science and Department of Physics
Massachusetts Institute of Technology
Cambridge, Massachusetts 02139

INTRODUCTION

Thirty five years of electron scattering have firmly established the electromagnetic probe as a powerful means for studying nucleons and nuclei. The clarity and precision of electromagnetic investigations have yielded discoveries central to our understanding of hadron and nuclear structure. For example, the demonstration of scaling in deep-inelastic electron-proton scattering essentially corresponded to "discovery" of the quark, and the most stringent tests of nuclear mean field theory, an approach which lies at the heart of much of our understanding of nuclear structure and dynamics, could be realized only with high resolution electron scattering. Focussing on nuclear physics, we can see that the great impact of the intermediate-energy, high intensity, ~1% duty factor electron accelerators commissioned in the 1970's (basically, the Saclay, MIT-Bates, and NIKHEF linacs) arose through a conjunction of new experimental possibilites with the emergence of a new set of ideas aimed at a microscopic foundation for nuclear models and at a quantitative understanding of nuclear forces in terms of mesonic degrees of freedom. Today, there is great optimism that a similar conjunction points to another major advance just ahead. The technology is now ready to provide two significant new capabilities for intermediate energy electron scattering: 100% duty factor (with high intensity) through CW acceleration (superconducting linacs, microtrons) or pulse stretching; full exploitation of spin observables, through polarized beams and internal targets. At the same time, the theoretical basis is being established for understanding how the color force becomes the strong force and how sub-nucleonic constituents determine nuclear properties. In these remarks, a few of the key areas currently thought likely to provide a focus for intermediate energy work in the 1990's will be outlined. These areas will come as no surprise, given the great number of workshops and project scientific justifications generated by the community over the last several years. However, given the limited time available, many important areas will go unmentioned; for example, the many new opportunities to be made available in the study of the structure of complex nuclei will not be covered (see the MIT-Bates and Illinois upgrade proposals for relevant discussions). Going beyond the discussion of experimental possibilities, some remarks will be made on the challenge to theoretical descriptions inherent in the developing program.

* This work is supported in part through funds provided by the U. S. Department of Energy (DOE) under contract DE-AC02-76ER03069.

STRUCTURE OF "ELEMENTARY" SYSTEMS

There is general agreement that a central question in intermediate energy nuclear physics is that of understanding the interplay between nucleon and nuclear degrees of freedom and the duality between hadronic and quark-gluon descriptions. It is crucial that the structure of the "elementary" A=1 and A=2 systems be understood as a foundation for building a more general picture of nuclear forces and structure. The importance of the new experimental capabilities proposed for the next decade is highlighted dramatically by the fact that we still do not have adequate data on the charge structure of the neutron, of the N→Δ transition, or of the deuteron. In the first two cases, the amplitudes are small, as anticipated by simple quark models; coincidence and/or spin measurements are needed to pull them out. The unit spin of the deuteron requires a spin measurement for separating the monopole and quadrupole charge distributions.

A. Two-Nucleon Bound State

The deuteron structure question is very special for nuclear physics. The two-nucleon problem provides the primary area for trying to understand the duality between the quark/gluon and baryon/meson picture, and the bound state multipoles, measured to large momentum transfer, would provide a complete characterization of the electromagnetic structure. Further, the isoscalar nature of the deuteron suggests that the physics beyond the traditional "nucleons-only" framework will not be dominated by simple pion exchange currents. A variety of predictions for the t_{20} observable[1] are shown in Figure 1. These predictions invoke different underlying degrees of freedom and, interestingly, lead to very divergent expectations at intermediate energies. Unravelling this situation by systematic examination of this and other intermediate energy nuclear observables is clearly part of the core program.

The theoretical situation also offers many opportunities for progress. A variety of quark and soliton models have been put forward for hadron structure and have, in many cases, been extended to nuclear structure and forces. None of the models claim to represent faithfully QCD dynamics in the confining regime; this is clearly not a task for which we are yet prepared. Instead, the models hope to capture essential elements of the underlying theory and to examine the consequences for nucleon and nuclear properties. The work of Lenz et al.[2] and of Gardner[3] are particularly relevant to the discussion of effective hadronic descriptions at low and intermediate energies. Lenz et al.[2] introduced a Hamiltonian quark model based upon saturation of color confining forces in color singlets. Basically, the confining forces act according to the spatial configuration of the multiquark system; the dynamics are driven solely by quark exchange between clusters. A remarkable feature of this "minimal" model is that it yields a considerable number of the qualitative phenomena seen in nuclear and low energy particle physics, including a hadron-hadron bound state with binding energy only a few

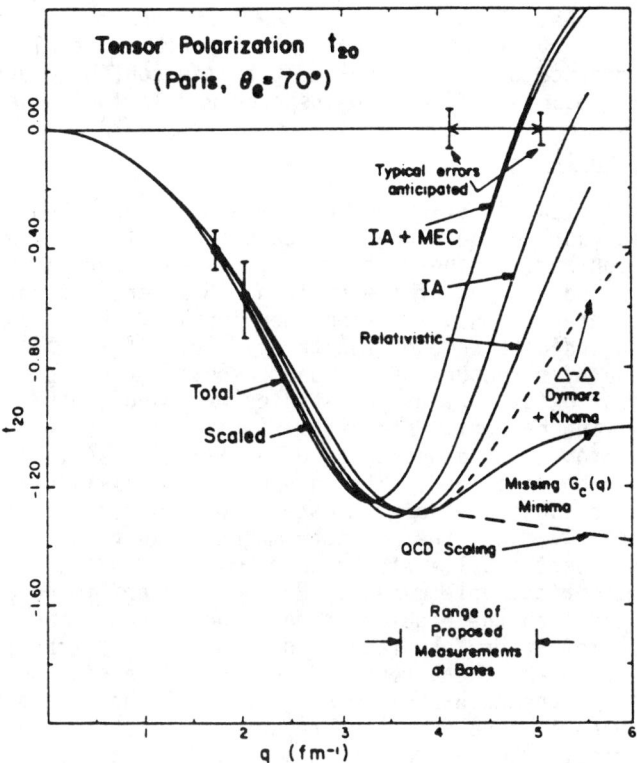

Fig. 1. Comparison between different theoretical predictions for T_{20} including impulse approximation (IA), meson exchange effects (MEC), relativistic corrections, Δ-Δ admixtures, QCD scaling, and the effect of phenomenologically filling in the IA minimum in $G_C(q)$. The data are from M. Schulze et al., Phys. Rev. Lett. **52**, 597 (1984). Figure taken from Reference 1.

percent of the hadron excitation energy scale. In the quark picture, the weak binding is understood as a consequence of the exchange process. The model also lends itself to reduction to an effective hadronic theory. It is clear that the reduction is not unique. Gardner[3] has examined the question of convergence of alternate effective hadronic theories to the bound state observables. She shows that the reduction involves extraction of both an effective hadron-hadron potential and, for the form factor, a consistent exchange-charge operator; the interaction and charge operator are coupled-channel operators describing the different hadronic internal states. She finds the remarkable result that with two rather different hadronic reductions, the quark model form factor is reproduced out to very large momentum transfer in both reductions with only the ground and first-excited hadronic states and with large, yet very different (i.e., an order of magnitude) exchange-charge contributions. This may not be the most encouraging result for explicit quark-effect searches. Nevertheless, it clearly points to the importance of proper treatment of the confining dynamics as one proceeds to more refined QCD-inspired models for nuclear dynamics.

B. Δ Properties

The nuclear interactions of the Δ have been studied extensively over the last decade, largely because the Δ is the lowest nucleon excitation and, in the quark model, has an internal structure closely related to that of the nucleon. However, before discussing its nuclear interactions, we stress how little is known about its structure and indicate some of the theoretical difficulties in addressing the issue of structure of a broad resonance; such difficulties must be addressed for many of the studies proposed in the region of the higher-lying baryonic "states".

The problem of understanding the N→Δ C2 transition multipole has already been mentioned. Its connection to possible Δ deformation has been discussed widely. Instead, we discuss briefly the diagonal $\Delta\gamma\Delta$ coupling, i.e., the Δ electromagnetic current. This has been studied experimentally with pion beams, the goal being extraction of the Δ magnetic dipole moment. However, we are immediately confronted with the problem of defining what we mean by the electromagnetic moment or form factor for a decaying "particle".

The problem can be resolved only by addressing the πN scattering dynamics in the Δ channel (more generally, the strong interaction decay channels coupled to the resonance). Within the standard $\Delta \leftrightarrow \pi$N isobar model, the πN transition matrix is given by

$$t_{\pi N}(E) = v_{\pi N\Delta} G_\Delta(E) v^+_{\pi N\Delta} \qquad (1)$$

$$G_\Delta(E)^{-1} = D(E) = E - m_\Delta - \Sigma_{\pi N}(E) = |D(E)| e^{-i\delta_{33}(E)} . \qquad (2)$$

The Δ propagator is modified by the πN self-energy $\Sigma_{\pi N}$, which both shifts the "mass" and provides the width. We introduce the quantity

$$Z = [D'(E_\Delta)]^{-1} \approx 1 - 0.4i . \qquad (3)$$

If the Δ were a true bound state, Z would represent the probability for the bound state to be a Δ. The significant imaginary part is indicative of the major role played by intermediate πN states in Δ propagation. The same physics must now be included in evaluating the Δ multipole moments, i.e., the photon couples not only to the bare Δ (which may, for example, represent a three-quark bag) but also to the intermediate pions and nucleons. For example, coupling to intermediate pions gives an effective contribution to the dipole moment

$$(\mu_\Delta^{++})_\pi \propto \frac{m_\Delta}{m_\pi} \left(\frac{Z-1}{Z}\right) \qquad (4)$$

which is both complex and energy-dependent. The full dynamical contributions[4] to the Δ^{++} magnetic dipole and electric quadrupole "moments" are shown in Figure 2; to provide reference, we note the

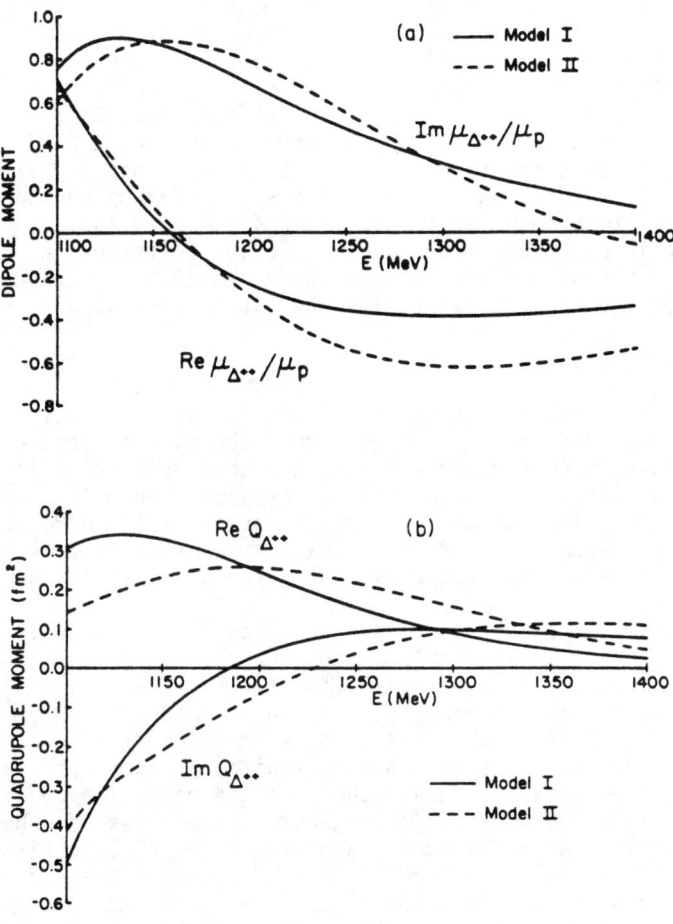

Fig. 2 Energy dependence of the effective (a) dipole and (b) quadrupole moments of the Δ^{++}. Taken from Reference 4.

SU(6) prediction $\mu_{\Delta^{++}} = 2\mu_p$ and the Isgur-Karl potential model prediction $Q_{\Delta^{++}} \approx -0.1$ fm^2. In the dipole moment, we see that the dynamical effects are appreciable and must be addressed in order to extract the physics, even when the experimental data on the resonance are completely adequate. For the quadrupole moment, the conclusion is even more severe, since the πN effects dominate the result expected from quark-quark interactions. Still, the analysis of such reactions is, in principle, straightforward, and quantities such as these are worth the effort. Indeed, it would be interesting to know whether a measurement of the Δ^+ dipole moment is feasible with CW, polarized electron beams. Obviously, the ratio of Δ^{++}/Δ^+ moments would provide a new test of the quark model.

Before leaving this section, we return to Figure 1 and note that the calculation of Dymarz and Khanna[5] is based upon a particular $\Delta\Delta$ probability in the deuteron ground state. The fact that the Δ is a broad resonance again complicates the interpretation. As stressed by Amado,[6] there is no straightforward probability interpretation for the Δ in a bound state. One may be able to get a qualitative handle on this question following the suggestion of Lee and Lipkin.[7] They stress that both the isospin ratios and longitudinal/transverse character in electropion production are very different for Δ electroexcitation and for knockout of a system having the Δ quantum numbers (formed through initial-state interactions).

C. Other Resonances

The Δ is, in the quark model, a simple "static" excitation of the nucleon. A deeper understanding of strong interaction dynamics demands that we understand the role of other degrees of freedom (orbital excitation, $q\bar{q}$ and gluon degrees of freedom) in the hadron spectrum. The importance (or unimportance!) of these degrees of freedom in, say, the first GeV of excitation should tell us much about QCD in the confining regime and provides a basis for studying the NN force.

Obviously, the odd-parity baryon resonances play a special role here, yet surprisingly little insight has been obtained so far into their structure.. A major reason for this is that the odd-parity N*'s are very broad (typically, $\Gamma \approx 150$ MeV) and strongly overlapping. On the other hand, the strangeness S = -1 resonances are rather narrow and thus provide a suitable system for initial studies. The radiative decay scheme of the neutral hyperons is shown in Figure 3. The electromagnetic decay rates have been calculated both within the bag model and within quark potential models.[8] The recent experiment of Bertini et al.[9] measured the electromagnetic decay width of the $\Lambda(1520)$ decay to the Σ and Λ ground states, γ_1 and γ_0, respectively. The interesting feature of the calculated widths is the sensitivity to the hadron wavefunctions. This is in sharp contrast for ground state decuplet→octet transitions; e.g., the ratio of widths

$$\frac{\Gamma_\gamma(\Sigma^\circ(3/2^+) \to \Sigma^\circ(1/2^+))}{\Gamma_\gamma(\Sigma^\circ(1/2^+) \to \Lambda(1/2^+))} \approx 3 \qquad (5)$$

$$\frac{\Gamma_\gamma(\Sigma^\circ(3/2^+) \to \Lambda(1/2^+))}{\Gamma_\gamma(\Sigma^\circ(1/2^+) \to \Lambda(1/2^+))} \approx 30 \qquad (6)$$

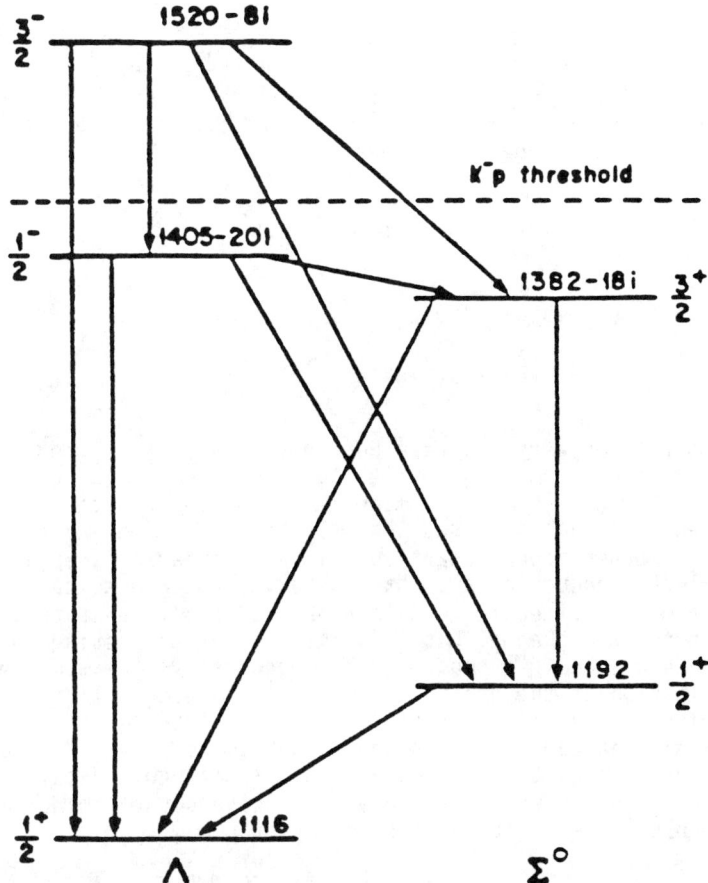

Fig. 3. Radiative-decay scheme for the low-lying neutral hyperons. Taken from Reference 8.

is practically the same for potential and bag model calculations and for calculations with or without explicit incorporation of strange quark mass effects. For the odd parity states, substantial multiplet mixing is expected because of the hyperfine interaction and because of the flavor symmetry breaking introduced via the strange quark mass. This multiplet mixing, in turn, strongly modifies the predicted radiative widths. For example, the q^3 bag model yields the ground state wavefunction

$$\Lambda(3/2^-) = 0.86\Lambda(1,3/2^-) + 0.35\Lambda(8,3/2^-)_{S=0}$$
$$- 0.35\Lambda(8,3/2^-)_{S=1} - 0.14\Lambda(8,1/2^-) \tag{7}$$

where the indices indicate the spin of the $(S1/2)^2$ quarks. The largest component is, as expected, the singlet. The ratio

$$\frac{\gamma_1}{\gamma_0} = \frac{\Gamma_\gamma(\Lambda(3/2^-) \to \Sigma(1/2^+))}{\Gamma_\gamma(\Lambda(3/2^-) \to \Lambda(1/2^+))} = 0.37 \qquad (8)$$

predicted with this wavefunction is changed by an order of magnitude if the octet admixtures are ignored:[8]

$$\left(\frac{\gamma_1}{\gamma_0}\right)\text{singlet only} = 3.8 . \qquad (9)$$

The experimental result[9]

$$\left(\frac{\gamma_1}{\gamma_0}\right)\text{expt} = 1.4 \pm .7 \qquad (10)$$

is not yet very accurate but certainly gives a good indication about the importance of studying these resonances.

How can the electromagnetic probe contribute to this program? Mecking[10] has stressed that electroproduction may compete favorably with other probes, particularly with a large acceptance detector. Electroproduction also has the advantage, in comparison with kaon beams, of directly reaching the $\Lambda(1405)$; this state is subthreshold in the K^-p channel but is particularly interesting in that it may have a large $q^4\bar{q}$ content. More generally, the study of electroexcitation of the proton to odd parity states will have a new impetus with the availability of intense CW beams, out-of-plane detectors, and polarized electrons and targets. Some may be studied through special signatures, such as the S_{11} through η decay. However, the full program will require an extensive set of coincidence measurements and extraction of spin observables.

NUCLEONS IN NUCLEI

The notion that nuclear wavefunctions are well represented by an antisymmetrized product of nucleon orbital wavefunctions defined self-consistently by a mean field has brought us a long way in our attempt to understand nuclear properties in terms of the NN force. Electron scattering has provided many of the most crucial quantitative tests of this idea. For example, the validity of a deformed intrinsic state in describing the rotation-vibration structure of rare-earth nuclei[11] is spectacularly tested by high resolution inelastic electron scattering; the notion that a central (i.e., angular momentum zero) orbit actually describes the probability distribution of the "last" proton in ^{206}Pb is beautifully confirmed by precise measurements of the ^{206}Pb-^{205}Tℓ charge distribution differences.[12] These are truly classic experiments in nuclear physics producing "textbook" results. However, these experiments, and a host of other elastic and inelastic electron scattering experiments, mainly probe the valence orbitals. Recent results in deep-inelastic electron scattering (i.e., energy transfer large compared to that needed to remove a proton) have reopened this fundamental question about the quantitative accuracy of mean-field descriptions. Clearly, these inclusive experiments sample all the nucleons in the nucleus.

The key observations have been those separating the quasielastic longitudinal and transverse inclusive response functions. Two results, generally confirmed by Saclay and Bates experiments, stand out. First, the longitudinal response function appears to be substantially smaller in the quasielastic peak region than expected on the basis of standard quasi-free calculations; this discrepancy appears to be very significant because, within the traditional framework of structureless nucleons, the total longitudinal strength obeys a simple charge sum rule (with corrections arising from correlations). Second, the transverse response function, while having approximately the expected magnitude in the quasielastic peak region, has substantial strength at the large energy-transfer side of the peak; this presumably reflects the importance of two-body currents. The clarity of the electromagnetic probe makes it difficult to ignore these inclusive observables. Most ideas put forward fall into one of two categories. Inspired by the EMC effect, one set of practitioners argues for "inflated" nucleons in the nuclear medium. This simple notion has the attractive feature of both decreasing the longitudinal strength and increasing the transverse/longitudinal ratio. A less provocative line of reasoning is that both the longitudinal and transverse "one-body strength" are spread by a variety of mechanisms (short-range correlations, final state interactions, ...) with the transverse strength greatly enhanced by multi-nucleon currents. Insight into these important questions has been and can be obtained in a variety of ways:

1. <u>Isospin Signatures</u>: The quasielastic response functions for the mirror nuclei ^{3}He and ^{3}H have been studied at Bates by a MIT-Saskatchewan-Virginia-NBS-Pittsburgh-WPI-Carnegie-Mellon-CEBAF[13] collaboration. The preliminary results are very intriguing. A comparison of the total longitudinal strength in the momentum transfer regime $q \approx 1.2 \text{fm}^{-1}$ with the ^{3}H and ^{3}He sum rule expectations of Schiavilla et al.[14] indicate that the significant short-range correlation effect expected for ^{3}He is confirmed. Recall that the sum rule depends only on the correlation structure of the ground state, which is calculable for realistic forces, and on the assumption of simple one body currents. On the other hand, the peak magnitude at $q = 2.5 \text{fm}^{-1}$ is significantly overestimated by the available calculations (none of which solve the continuum three-body problem, as is required for a fully consistent evaluation of the <u>distribution</u> of strength).

The transverse response functions are also interesting. For example at $q = 2.5 \text{fm}^{-1}$, the ratio of ^{3}H and ^{3}He peak height is consistent with the expectation based upon one-body currents. However, the response functions become equal away from the quasielastic peak, suggesting dominance by two-body "quasi-deuteron" currents.

2. <u>y-Scaling</u>: At very large momentum transfers, the quasielastic response function will scale in the variable y, which represents the

parallel component of the struck nucleon momentum (as calculated under the assumption of quasifree kinematics). The SLAC results of Sick et al.[15] for ^{3}He are well known and seem to indicate scaling even for values of $|y|$ well beyond the "Fermi-momentum". There is some controversy over the question of whether or not the scaling function is providing an accurate measure of the momentum distribution.

The uncertainty arises from two sources. First, within a "nucleons-only" framework, hard-core correlations or final state interactions with a constant mean free path may preserve scaling but alter the scaling function. Second, the neglect of internal nucleon structure in the derivation of y-scaling may be questioned at very large momentum transfer, even though the energy transfer is small.

The latter question has been investigated by Kumano[16] in the quark exchange model discussed earlier. He considers a pair of composite hadrons bound in an overall mean field potential, with the hadron-hadron interactions governed entirely by quark exchange. He evaluates the response function for large negative y, both in the quark model and with an "equivalent" model consisting only of ground-state hadrons interacting through an effective potential which reproduces rather well the free-space scattering phase shift. The effective potential is strongly repulsive at ranges characteristic of hadron size. The remarkable result is that the two response functions differ by at most of a factor of two in the large negative y region even when the response function has fallen by many orders of magnitude from the quasielastic peak value (i.e., y = o). Consequently, the scaling phenomenon appears to survive in the face of hadron compositeness for energy transfers much smaller than those corresponding to quasifree knockout. While this is encouraging for attempts to extract the nucleon momentum distribution well beyond the Fermi surface, it reinforces the elusiveness of explicit signatures of quark substructure in low energy observables.

3. (e,e'p): Coincidence studies are clearly of paramount importance for understanding the basic issues raised by the inclusive results. Although a "complete" experimental program awaits high intensity, CW electron beams, extremely interesting exploratory studies have been carried out at Saclay, NIKHEF and Bates. The original Saclay work was largely motivated by spectroscopy. The energies and momentum distributions for nucleon orbitals were extracted. The recent work has, in addition, explored the issues of nucleon structure in the medium and of the underlying reaction mechanism by carrying out longitudinal/transverse separations, by mapping out the momentum transfer dependence of the process, and by exploring the large missing energy regime. Recent Bates data for ^{12}C(e,e'p) indicate that the longitudinal/transverse ratio at q = 400 MeV/c for p-shell knockout is consistent with that for a free nucleon,[17] as is the q-dependence over the range 400 to 800 MeV/c.[18] On the other hand, large strength is seen for a variety of electron kinematics in the missing energy regime at and beyond s-shell knockout. An example is shown in Figure 4, where the electron kinematics correspond to the large energy loss side of the quasielastic peak.[19] The dominance by large missing energy strength is qualitatively con-

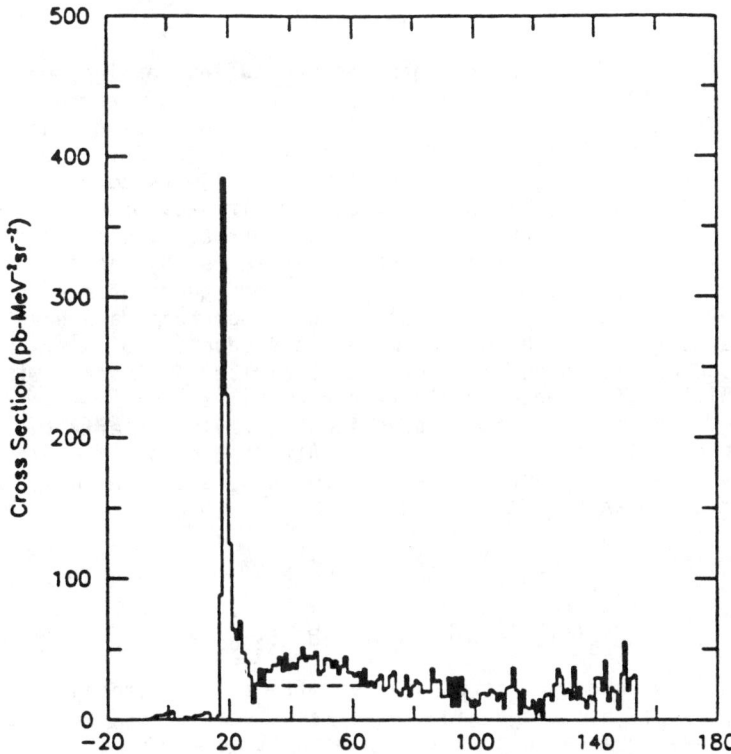

Fig. 4. Missing energy spectrum for $^{12}C(e,e'p)$ in the "dip" region $\bar{q}$ = 400 MeV/c, ω = 200 MeV. Taken from Reference 19.

sistent with the quasi-deuteron dominance suggested by the ^{3}H - ^{3}He transverse response functions. This is a good example of how the exclusive and inclusive studies combine to provide insight about the underlying reaction mechanism. A more quantitative understanding awaits a more extensive experimental program and the development of appropriate theoretical tools. Clearly, the future experimental program will have to include more angular coverage of the (e,e'p) final state, neutron detection, out-of-plane measurements, and studies using polarized electrons. All of these initiatives are in the early stages of implementation at various laboratories and form a core subject for future CW programs. However, the optimism that these experimental studies will settle fundamental questions in intermediate energy physics must be tempered by the reality that needed theoretical developments are lagging behind. Obviously, coincident hadron detection by definition puts a higher premium on analysis in order to get out the physics. Beyond this, technically feasible approaches based upon an acceptable theoretical framework have not been developed yet for treating large missing energy processes. The community must find a way of fostering a theoretical base commensurate with the anticipated experimental developments.

Δ IN NUCLEI

The study of excited nucleons in the nuclear environment has been one of the distinguishing features of intermediate energy physics. In particular, the availability of intense pion beams has allowed a systematic study of Δ-nucleus interactions. The electromagnetic probe has several features which allow us to get at complementary information: the photon (real or virtual) probes the entire nuclear volume, in contrast to the surface reactions induced by pions; with electrons, the momentum transferred by the photon can be varied independently of the energy transfer; the photon energy can be lowered through pion threshold. Some important data have been obtained already in the electromagnetic production of Δ's. However, the features discussed above will be exploited fully only with the availability of intense CW electron beams.

To frame the following discussion, we review briefly the results of Δ-hole analyses of pion elastic scattering data. The pion-nucleus transition matrix is expressed in a manner which is formally similar to Equations (1) and (2):

$$T_{\pi\pi}(E) = V_{\pi N\Delta} G_{\Delta h}(E) V^+_{\pi N\Delta}$$

$$G_{\Delta h}(E)^{-1} = E - M_\Delta - \Sigma_{\pi N}(E) - H_\Delta - \delta w - w_0 - V_\Delta \qquad (11)$$

where H_Δ, δw, and w_0 incorporate the effects of Δ propagation and binding, Pauli blocking, and pion elastic scattering, respectively. All of these are calculable within a shell model framework. The last term, V_Δ, represents the Δ-nucleus interaction and is the quantity of interest. Writing

$$V_\Delta(r) = V_C \frac{\rho(r)}{\rho(o)} + V_{LS} f(r) 2\vec{L}_\Delta \cdot \vec{\Sigma}_\Delta \qquad (12)$$

with $\rho(r)$ the nuclear density and $f(r)$ a surface peaked function, the interaction parameters extracted from pion elastic scattering are[20]

$$V_C \approx (20 - 45i) \text{MeV} \qquad (13)$$

$$V_{LS} \approx (-10 - 4i) \text{MeV} \qquad (14)$$

The striking result is the large central imaginary part, indicative of a strong damping mechanism arising from annihilation ΔN→NN. One thrust of the complementary electromagnetic studies is testing of this phenomenology of Δ-nucleus interactions. Another is in helping to elucidate the reaction processes underlying the phenomenology, particularly since pion absorption studies indicate that a simple two-body ΔN process may not be sufficient for explaining much of the absorption strength.

Conceptually, the simplest electromagnetic process to discuss is the total photon cross section. Because of the weak photon absorption, this inclusive process essentially displays the Δ shape in the medium plus "background" processes. Unfortunately, the data are not so easily obtained; the Bonn data [21] shown in Figure 5 include a

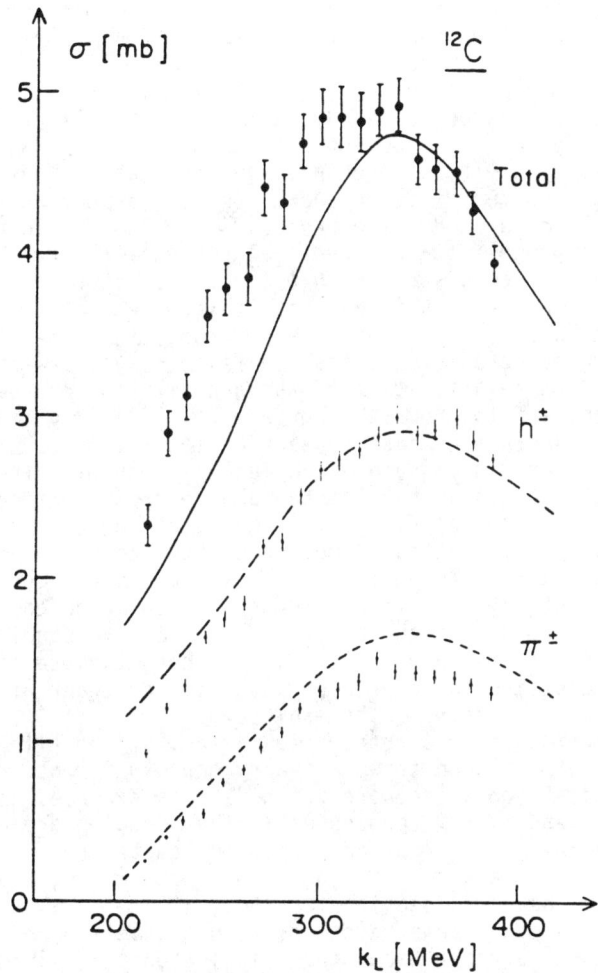

Fig. 5. Total photoabsorption cross section on ^{12}C. The observed charged pion ($\pi^\pm$) and charged hadron ($h^\pm$) cross sections and the extrapolated data for the total cross section are taken from Reference 21. The curves result from Δ-hole calculations of Reference 22.

significant contribution from neutral particles, which are included by Monte Carlo modelling. The Δ-hole calculations, constrained by the pion data, including partial cross sections for charged hadrons and charged pions, are also shown.[22] Although the measured charged particle yields are well reproduced, there is significant extra strength in the total cross section for energies below the peak. Further investigations of this strength rely upon more exclusive processes.

The (e,e'p), (e,e'π), (e,e'Nπ), and (e,e'NN) processes are obvious candidates for study, all of which will profit from CW beams. The proton knockout process has been explored both with real and with virtual photons. In both cases, the inclusive proton spectra show two distinct peaks, one corresponding to the nucleon from Δ→Nπ decay and one corresponding to the nucleons from the annihilation process ΔN→NN. For real photons, the energy dependence of the peaks is consistent with a Δ-doorway hypothesis.[23] The (e,e'p) reaction appears consistent with this when the momentum and energy transfer are fixed to quasifree Δ-production kinematics. However, when the electron kinematics are fixed on the low-energy side of the Δ peak, the relative size of the annihilation to quasifree decay peaks appears to grow substantially.[24] This may signal the importance of another competing reaction mechanism (e.g., direct NN emission) and is certainly not inconsistent with the previously discussed inclusive data (total photon cross sections, inclusive ^{3}H-^{3}He response functions, (e,e'p) "dip" region missing-energy spectra). Nevertheless, whether or not these exploratory coincidence studies are defining a consistent phenemonological framework for understanding the nuclear response to large energy transfer, the basic point is that the extensive coincidence studies permitted by CW beams will be very powerful. Further, the use of spin observables sensitive to interferences of the dominant amplitude with "background" remains to be exploited.

Reactions to discrete nuclear final states allow a particularly crisp confrontation between theory and experiment. That is, the theoretical tools are more refined. For example, the coherent (γ,π°)[25] and (γ,γ)[26] processes in the Δ region are amenable to a unified Δ-hole description with pion scattering. Basically, different matrix elements of the Δ-hole propagator, Equation (11), are probed. The cross section for ^{4}He(γ,π°)^{4}He is shown in Figure 6a. The Δ-hole model constrained by pion scattering appears to do very well. On the other hand, the existing data really only probe the maximum of the distribution. With CW beams, two orders of magnitude of cross section are still available with existing detector technology. For example, ^{12}C(γ,π°)^{12}C data are shown in Figure 6(b) on a logarithmic scale. Clearly, CW beams will open a new regime for study, much more sensitive to the underlying dynamics.

Many other processes will be investigated. For example, Compton scattering is more sensitive to the spin-dependence of the Δ-nucleus interaction. Charged pion electroproduction to discrete nuclear states will allow a "filtering" of various reaction terms; e.g., the Δ production piece is dominated by spin-non-flip terms, while the background is primarily spin-flip. With the appropriate choice of

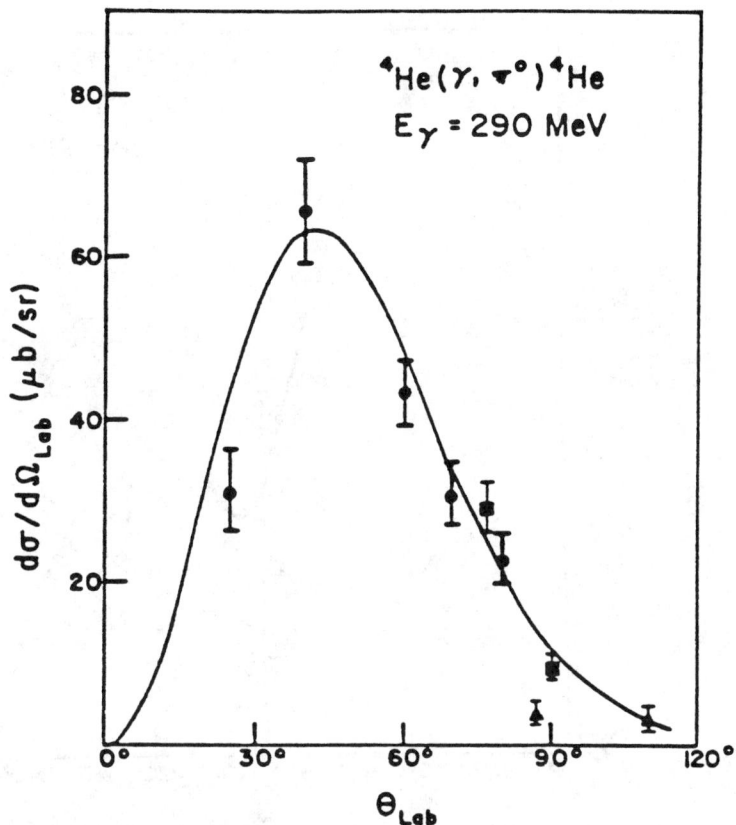

Fig. 6a. Coherent (γ,π°) cross sections (a) ^{4}He at 290 MeV; data from Reference 25; (b) ^{12}C at 242 MeV; data: solid points from J. Arends et al., Z. Phys. A311, 367 (1983); crosses from J. Comuzzi, MIT Ph.D. thesis (1983), unpublished. The curves are Δ-hole results from J. H. Koch and E. J. Moniz, Phys. Rev. C27, 434 (1983).

quantum numbers, it will be possible to restrict the contributions to electroproduction and, in principle, to map out the interior pion wavefunction.

Finally, we note that the in-medium study of higher resonances is likely to be very interesting. As already mentioned, the odd-parity resonances look particularly interesting. Naive arguments about the size of such resonances indicate that their propagation may be strongly damped in the medium. This is certainly an area needing both theoretical study and a phenomenological study of accessible signatures for the interesting physics.

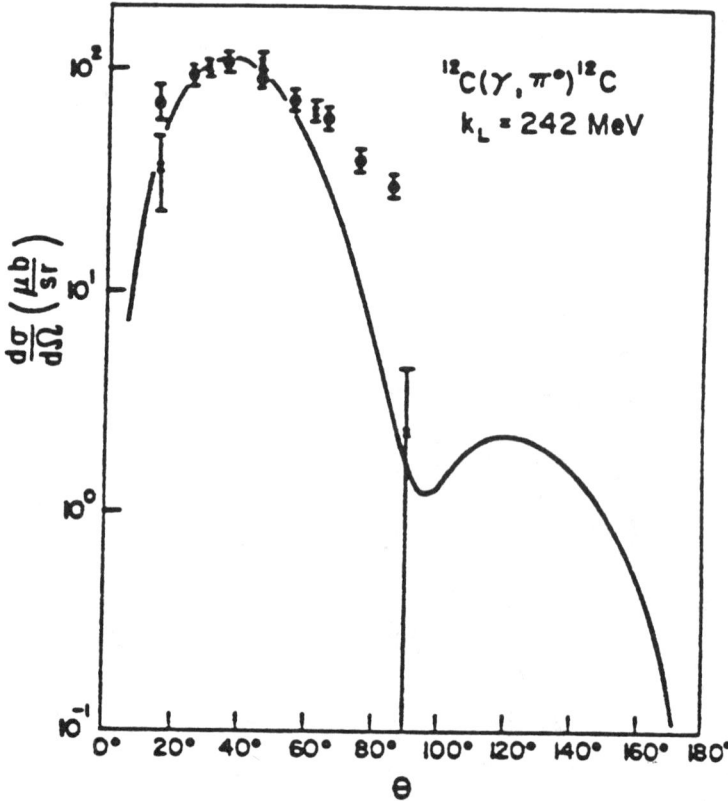

Figure 6b.

CONCLUDING REMARKS

Inclusive electron scattering has made an enormous contribution to our understanding of hadron and of nuclear structure and to defining the questions which are driving the field in new directions. With intense CW intermediate energy electron beams and with the opportunity to exploit spin observables, central contributions to many of the most crucial questions are anticipated. A few examples have been discussed. However, this progress hinges much more strongly than did the earlier one on the development of appropriate theoretical frameworks for isolating the interesting physics.

REFERENCES

1. L. Antonuk et al., Bates Linear Accelerator Center Annual Report (1985) pg. 33.
2. F. Lenz et al., Ann. Phys. $\underline{170}$, 65 (1986).
3. S. Gardner, to be published.
4. L. Heller et al., Phys. Rev. $\underline{C35}$, 719 (1987).
5. R. Dymarz and F. Khanna, Phys. Rev. Lett. $\underline{56}$, 1448 (1986).
6. R. D. Amado, Phys. Rev. $\underline{C19}$, 1095 (1979).
7. H. J. Lipkin and T.-S. H. Lee, Phys. Lett. $\underline{B183}$, 22 (1987).
8. E. Kaxiras et al., Phys. Rev. $\underline{D32}$, 695 (1985).
9. R. Bertini et al., preprint.
10. B. Mecking, private communication.
11. W. Hersman et al., Phys. Rev. $\underline{C33}$, 1905 (1986).
12. B. Frois et al., Nucl. Phys. $\underline{A396}$, 409c (1983).
13. D. Beck et al., Bates Linear Accelerator Annual Report (1985), pg. 69, to be published.
14. R. Schiavilla et al., Illinois preprint ILL-(NU)-86-#60.
15. I. Sick, D. Day and J. S. McCarthy, Phys. Rev. Lett. $\underline{45}$, 871 (1980).
16. S. Kumano, MIT Ph.D. thesis (1985); to be published.
17. P. Ulmer et al., to be published; for a preliminary account, see H. Baghaei et al., Bates Linear Accelerator Center Annual Report (1985) pg. 14.
18. L. Weinstein et al., to be published; for a preliminary account, see H. Baghaei et al., Bates Linear Accelerator Center Annual Report (1985) pg. 14.
19. R. Lourie, et al., Phys. Rev. Lett. $\underline{56}$, 2364 (1986).
20. For a review, see E. J. Moniz, in "Symmetries in Nuclear Structure", edited by K. Abrahams et al. (Plenum, 1983) pg. 251.
21. H. Rost. Bonn Report IR-80-10 (1980).
22. J. H. Koch et al., Ann. Phys. $\underline{154}$, 99 (1984).
23. S. Homma et al., Nucl. Phys. $\underline{A446}$, 341c (1985).
24. H. Baghaei et al., to be published; for a preliminary account, see H. Baghaei et al., Bates Linear Accelerator Center Annual Report (1985) pg. 14.
25. D. R. Tieger, et al., Phys. Rev. Lett. $\underline{53}$, 755 (1984).
26. J. Miller et al., to be published.

THE FUTURE OF HIGH ENERGY ELECTRON BEAM PHYSICS

James D. Bjorken
Fermi National Accelerator Laboratory
Batavia, Illinois 60510

I. INTRODUCTION

It is a pleasure for me to be here to help celebrate this occasion, especially since so many of my own years in physics have been dependent upon the development of high energy electron beams as probes of the structure of matter. This talk will not attempt to be a thorough review of what has been and is being done. Instead, all I can offer is a quite personal view of the future importance of electron-beam work.

A. The Electron as a Probe of Matter

We all know the long history of the electron as an especially useful probe of matter, with its usefulness enhanced because its main interaction is only electromagnetic. Electron collisions leading to excited levels in atoms, molecules, and nuclei provided a rich source of information about their internal structure. For relatively gentle collision processes, these processes contributed to spectroscopy. For violent processes - those for which the energy transfer from electron to target is large compared to the level spacing - an "impulse approximation" typically is appropriate. This implies that the electron interacts with the constituents as if they were quasi-free; their initial state wave-function provides an initial-state momentum distribution, while the final state of the struck object can be considered as a plane wave (better, a wave packet $e^{iqx}\psi_0(x)$, provided a coarse-grained average on final-state energies is admitted).

Nonrelativistic kinematics for the target system suffices for the study of molecules, atoms, and nuclei. However, this situation changed when electrons were used to study the proton. The traditional frame of reference, in which the target is initially at rest, became awkward because the quark constituents manifestly could not be handled norelativistically. The breakthrough came in going to the opposite extreme and choosing a center of mass frame of reference in which the target nucleon is extreme-relativistic. In such a frame, the collision time is arguably very small because of Lorentz-contraction of the target. In addition internal motions (clocks) are slowed down by relativistic time-dilation. The result is that the old impulse approximation works. Electron-scattering from quark constituents happily became a viable concept, with a remarkably simple descriptive structure (the parton model) emerging.

© American Institute of Physics 1987

The revelation of quarks within the nucleon via electron scattering is now about to undergo the next step. Half the momentum of an extreme-relativistic proton is not carried by quarks, but rather by electrically neutral gluons. This component is now seen quite clearly in the high energy proton-antiproton collisions at CERN. As far as electron interactions are concerned, the gluon component should be seen clearly at the HERA electron-proton collider under construction at DESY. The process is simply (internal) gluon bremsstrahlung of the struck quark; the gluon should be seen as a jet of hadrons.

But we are getting ahead of the story. It is only to say that electron interactions will still be probing the constitution of matter, possibly beyond the quark level, in the future. But the future prospects in this direction are balanced by attractive prospects in the other directions in which electron-beam physics has evolved. In particular, the electron is a carrier of photons. Also the electron serves as a probe of the vacuum (in the sense of electron-positron colliders).

B. The Electron as a Source of Photons

Perhaps the most important category here is synchrotron radiation; however this supremely practical and important application of electron storage rings is beyond the scope of what we try to cover here.

But at high energy, the electron is not only electron but also photon; its Lorentz-contracted coulomb field becomes a beam of photons, according to the Weizsacker-Williams picture. Indeed, this is a primitive example of the parton model in action. These photons can at high energy in turn be viewed as having constituents, e.g. e+e- or other lepton pairs. The Bethe-Heitler processes of pair production or bremsstrahlung are just the liberation of these constituents by a collision with a target.

Especially interesting are the hadronic constituents of the photon - the vector mesons ρ, ω, ϕ as well as quark-antiquark parton pairs. These all can be liberated in collisions by a process analogous to the QED Bethe-Heitler processes. However, we have already described such processes from the parton point of view. In a center-of-mass frame, this process was viewed as the electron probing the constituents of the target nucleon or nucleus. In the laboratory frame the same constituents were viewed as constituents of the electron. So where are these constituents? Are they in the electron or in the nucleon? - or in both? - or in neither? The answer is yes. In the center-of-mass parton picture, the hadron structure which is probed in this "vector-dominance" limit is that of the "wee" parton component; each parton carries a very small fraction of the total momentum. Therefore it is poorly localized and resides in a longitudinal region about one fermi thick, very thick relative to the

Lorentz-contracted pancake containing the baryon-number. Thus in the laboratory frame this region becomes very long, of order γ fermi, where γ is the Lorentz-boost factor of the hadron from the center-of-mass frame to the laboratory. This long longitudinal coherence length is fundamental. What is in a real sense probed are the contents of the vacuum, as disturbed by the passage of hadron and electron through it: fluctuations containing charge and/or color which are able to interact with the passing electron and/or hadron. In any case, this disturbed vacuum can, under appropriate kinematical circumstances, be regarded as "belonging" to one or the other probes.

Consequently high energy photoproduction studies have yielded a rich harvest of information on particles such as those containing strange and charmed quarks which are not obvious constituents of the proton. And they have given very persuasive evidence for the "vector-dominance" ideas and the existence of long longitudinal coherence lengths. The dynamics is subtle; much has been learned but there is still a long way to go.

C. The Electron as a Probe of Vacuum

The most direct way to use electrons to study the properties of the vacuum is via e+e- annihilation. From the point of view of the old Dirac hole theory, the electron scatters from a negative energy constituent of the vacuum, and thereafter drops into a negative energy state (the positron "hole") in order to satisfy energy-momentum conservation. This process has produced a tunable probe of all kinds of vacuum excitations, with splendid signals at the frequencies appropriate to ρ, ω, ϕ, ψ, and Υ. One can dial to one's favorite quark. We hope this soon will include the top quark. One can produce anything with charge, such as the τ lepton. In addition, one soon will dial to the neutral weak boson Z^0, which should yield a large number of interesting decay products - maybe even including some information on the Higgs sector. We need not belabor the point here: history has provided ample evidence of the outstanding successes of this technique.

II. CURRENT PROBLEMS AND THEIR STATUS

In assessing the status of electron physics, we may use again the previous categories:

A'. The Electron as a Probe of Matter

At present the short-distance hadron structure is being probed most incisively by neutrinos and muons; the latter in my opinion is the probe of choice at least until HERA is commissioned. A central issue of hadron structure, however lies in the gluonic component of the nucleon. Here a probe by leptons is at best indirect. The splendid successes of the UA1 analysis of two-jet and three-jet events argue for the reality of hard gluon-gluon collisions. And the energy scale probed already competes favorably with what can be reached by a 100 x 100 GeV upgraded LEP electron-positron collider.

Nevertheless there are outstanding issues for lepton-hadron interactions. The structure function at small values of the scaling variable x (we are thinking of $x \lesssim .03$) especially need clarification: what is the Regge behavior both for "singlet" (vacuum trajectory) and "non-singlet" exchanges? What is the Q^2 dependence? And what is the A dependence? All these problems are subtle, and will help to reveal the nature of the vacuum "cloud" which couples to energetic hadrons.

B'. The Electron as a Source of Photons

Present work on photoproduction of charmed hadrons is very fruitful. About 1% of the total hadroproduction cross section leads to charmed-particle final states. A recent experiment at Fermilab (E691) has collected over a million events containing charmed hadrons; already analysis of these data has provided accurate values of charmed-meson lifetimes. Event samples in specific decay channels easily compete with the best from e+e- experiments. The key to this success has been the development of silicon microstrip detectors which are able to reconstruct the secondary decay vertices of the charmed hadrons, thereby removing most of the copious backgrounds which has inhibited progress up to now. Prospects for extending this success to bottom-quark photoproduction appear to be bright, and a new Fermilab experiment (E687) designed for photon energies of about 300 GeV should be quite productive.

At lower energies there is also much activity. Photoproduction of ψ and χ charmonium states, including A-dependence, teaches much about strong-interaction production mechanisms, and again about longitudinal-coherence phenomena. And we even see a resurgence of interest in photoproduction from nuclei in the 10-15 GeV range utilizing gas-jet targets in electron (or positron) storage rings. And at somewhat lower energies there is sure to be much interest in this subject with the intense, high quality photon beams at CEBAF.

C'. The Electron as a Probe of Vacuum

Electron-positron annihilation continues to provide rich yields of physics over a broad range of energies. The spectroscopy of ψ and ψ' decays is still providing important data, e.g. the mysterious state at 2.2 GeV seen at SPEAR. At charm threshold we have hints of either $\Delta C = 2$ weak charm decays or of unexpectedly large D-D mixing or both. At PEP energies, there have occurred accurate measurement of B-meson lifetimes, and at PETRA a handful of puzzling events containing muons, seen only at the highest energies, invite follow-up measurements at TRISTAN, the higher energy ring now being commissioned in Japan. The threshold region for bottom-quark production remains very rich: the spectroscopy of upsilon states and the detection of bottom mesons are at present dominated by results from the e+e- machines at Cornell and DESY.

Finally there is the physics of Z^o decays expected from SLC and LEP. Everything with electroweak coupling, hopefully including top quarks, Higgs bosons, and exotica, should emerge as decay products - provided their mass is not too large. In addition the increased yield of charm and bottom quarks (a factor 10-100 over PEP/PETRA) provides new opportunities for more sensitive study of these systems. A measure of success for e+e- physics in general is that two thousand PhD physicists have each chosen to devote several years to this single experiment (the properties of Z^o decays) at a capital cost of at least one billion dollars. We can only hope that nature is especially generous in rewarding such an enormous effort.

III. FUTURE PROSPECTS

In looking at the future, we again make this division of subject matter:

A". The Electron as a Probe of Matter

As we already mentioned, the classic problem of probing the constituents of the nucleon is rapidly becoming obsolete: the problem is now the constitution of the quark. Indeed the question of compositeness of quark and of electron stand on the same footing. And in fact one can generalize this to the whole periodic table of building blocks. Probably a majority of theorists would agree that internal structure, if any, of the electron neutrino and of the top quark is the same problem as understanding of internal structure of the electron and of ordinary quarks. Thus the question of the structure of matter has become more detached from that of the structure of the nucleon. Hence lepton-nucleon scattering may in the future play less of a special role that it has in the past.

The electron-proton collider at HERA will be a watershed in this regard. At 30 GeV on 800 GeV it boasts a center-of-mass collision energy which is high compared to e+e- colliders, but low compared to present proton-antiproton colliders. The collisions are probably "cleaner", i.e. more easily able to interpret, than hadron-hadron collisions, but "dirtier" than the pure e+e- collisions. The rate of events of interest in higher than e+e-, lower than $p\bar{p}$. Thus this ep collider interpolates between the extrema of e+e- and $p\bar{p}$ physics. Does this mean it is better to work at the extrema? Or do ep collisions optimize the advantages of the "extremal" processes and minimize their disadvantages? With HERA we should find out.

B". The Electron as a Source of Photons

The present success in photoproduction experiments suggests a long future ahead for them, at low energy and high, in collider as well as fixed target. One must remember that HERA represents a rich source of photon-proton collisions, although tagging the secondary low Q^2 electron may be difficult. At a lower energy scale, the processes of heavy quark production stand out, in my opinion, as a field with great longevity. At still lower energies, photon-photon collisions at PEP hold out promise for continuing rich returns, both for meson spectroscopy and for hard-collision studies. And at still more modest energies the burgeoning interest of nuclear physicists in photoproduction processes at PEP and at CEBAF guarantee a considerable future.

C". The Electron as a Probe of Vacuum

We have already implied that this is a very central issue for the future, and that e+e- colliders represents the most attractive technique. Of course the push toward higher energy will continue. TRISTAN (30 x 30 GeV) is being commissioned. SLAC's linear collider is not far behind, and LEP should follow in 1989. But in parallel with this should also come a push toward higher luminosity and detection capability at all energy scales, especially landmark scales. This includes charm and bottom threshold regions, especially charm, where one detector, Mark III operating only sporadically at SPEAR, can scarcely do justice to the physics potential present there. The Beijing IHEP machine should help. But is it unthinkable to push for a luminosity of 10^{33} or above in that energy region?

IV. THE FAR FUTURE

The push to the highest center-of-mass energies using electrons has two branches. One of these could be a byproduct of future multi-TeV hadron colliders such as the SSC or CERN's LHC. An electron ring of modest (30-100 GeV?) energy appended to such a machine could lead to ep collisions with center-of-mass energy very high compared to HERA. The attractiveness of this option will depend to a considerable degree on HERA's productivity.

The main perceived branch is in e+e- linear colliders. Here the technology is daunting. One needs a very high accelerating gradient and high efficiency; even so the power bill will be a difficulty. Not only is the energy scale hard to attain, but the requirements on luminosity are extreme. Submicron size beams at the collision point must be provided, leading to severe demands on precision alignment, beam stability and beam quality.

The energy scale of interest for future e+e- machines may best be dictated by the hadron-hadron colliders. Already CERN's p$\bar{\text{p}}$ collider has begun to explore quantitatively the mass region of 100-200 GeV, available only to LEP II. Fermilab's Tevatron will go well beyond that. It may well be that new phenomena point to a specific mass scale, thereby inviting the appropriate e+e- initiative. The obvious example would be observation of a heavier Z' boson possessing an appreciable width into e+e-. But less obvious cases can be entertained as well.

At lower energy scales, we have already pleaded the case for higher luminosity at landmark energy scales. Additional landmarks may proliferate even at the soon-to-be-attained energy scales, for example at top-antitop threshold, wherever that may be. And the frontiers may push to lower mass-scales. For example, a new initiative for measuring the gyromagnetic ratio of the muon is underway. However, for the result to be useful the hadronic vacuum-polarization correction to the $\alpha/2\pi$ Feynman diagram needs to be known to better accuracy. This can be done experimentally, provided the e+e- cross-section to hadrons can be measured to better than one percent from threshold to the 1 GeV mass region. The Soviet machine VEPP-II may be able to do this.

As another extreme example of a low mass frontier, even now some people entertain the idea of a high luminosity ϕ factory to make a large number of $K_S K_L$ pairs for CP violation studies. A 500 x 500 MeV e+e- collider with luminosity in excess of 10^{33} cm^{-2} sec^{-1} would be required. There are alternatives, e.g. photoproduction at high luminosity (and good duty factor) at any energy above a few GeV (CEBAF?) If one million <u>hadronic</u> photoproduction events per second could be attained (and Bethe-Heitler e+e- pairs could be tolerated), over 10^{10} ϕ's could be produced per experiment.

Perhaps the ideal ϕ factory might be made with a 1.02 TeV positron beam incident on a fixed target, in particular the atomic electrons therein. 10^{10} positrons per second onto a 15 gm/cm^2 H$_2$ target would yield a luminosity of 10^{35} cm^{-2} sec^{-1}. While no one would propose a facility dedicated to this purpose, it might be a useful zeroth generation experiment for a 501 x 501 GeV linear collider, provided positrons could be accelerated in the reverse direction through the electron arm. The specifications of beam quality for the resonant ϕ production experiment would be negligible relative to the collider specifications. So during the considerable tuneup time such a collider would inevitably require, very useful physics could be carried out.

While this example of ϕ factories may be a fanciful one, it does serve to exemplify how electron physics, despite its flourishing over many orders of magnitudes of increase in beam energy since the pioneering experiments 35 years ago here, still in principle retains its unity of purpose.

Symposium - 35 Years of Electron Scattering

J. D. Walecka
The Continuous Electron Beam Accelerator Facility
Newport News, Virginia

Concluding Remarks

When I was coming here on the plane the other night, I did not have the foggiest idea of what I was going to say. I went for a long walk last night, and now I find I can talk for at least an hour. But I am not going to. I enjoyed the symposium. I particularly enjoyed yesterday because it took an historical approach to the subject. We are honoring 35 years of electron scattering, but in fact, Professor Hanson went back to Rutherford. We started with, if I recall one of his slides correctly, something like a 100 KeV gun and detector, and we ended last night with a 100 GeV beam and the L3 detector. And that is all within one lifetime. It is really an impressive history.

Since we are talking about history, let me talk a little bit about history myself. I first got involved in electron scattering through a talk Bob Hofstadter gave when I was a postdoc at CERN in 1958. I was impressed by the quality of the experiments, and particularly by the interaction between theory and experiment. He would present this nice theoretical cross section, the Rosenbluth cross section, the Jankus formula, etc., and from that and the experiments he would then deduce all this marvelous information about what nucleons look like, what the deuteron looks like, what nuclei look like . . . It is a nice field. It really is high quality information; you know what you measure. I went to Stanford and was associated with Stanford and HEPL for the 20 year period from 1959 to 1979; it was really a very exciting time and a very exciting place. It was built on the klystron, which was developed at Stanford, and on the electron linac. If you stood in the middle of HEPL you could see Hofstadter's spectrometers, and you could see the storage rings. I remember Burt Richter, Gerry O'Neill, Dave Ritson, Bernie Gittleman, and Carl Barber, night-after-night, trying to make those storage rings work. Of course, HEPL led to SLAC, SPEAR, and PEP. The first superconducting cavities were developed there; John Pierce built the first one, if I am not mistaken. Large-scale refrigeration was developed there at Stanford. The first free electron laser was down in the basement of HEPL. it was really a marvelous time and a marvelous place.

I also know bj from Stanford. In fact we overlapped during that period. He and I shared an office for a good fraction of that period. I do not want to embarrass bj, but he has always been my model of a physicist. He can do theoretical physics at the forefront; he can do it with the best of them. But he also realizes physics is an experimental science, and he works closely with experiment. To me, that was one of the things that made Stanford a special place during all that time. It was the close interaction between theory and experiment. After bj's talk about high energy physics, I have nothing to add to that subject. I am therefore going to concentrate on nuclear physics.

Let me go back to the beginning. Why do we do nuclear physics? First the nucleus is a unique form of matter. It consists of many baryons in close proximity. Second, all the forces of nature are present in the nucleus - strong, electromagnetic and weak. The nucleus provides a unique microscopic laboratory to test the structure of the fundamental interactions. Furthermore, the nuclear many-body problem is of intrinsic intellectual interest. In addition, most of the mass and energy in the universe around us comes from nuclei and nuclear reactions. Finally, in sum, nuclear physics is the study of the structure of matter.

Why do we do electron scattering [1]? First, the interaction is known. It is governed by quantum electrodynamics (QED), which is the most accurate physical theory we have. Second, the electron provides a clean probe; we know what we measure. In addition, the interaction is relatively weak, so we can make measurements without greatly disturbing the structure of the target.

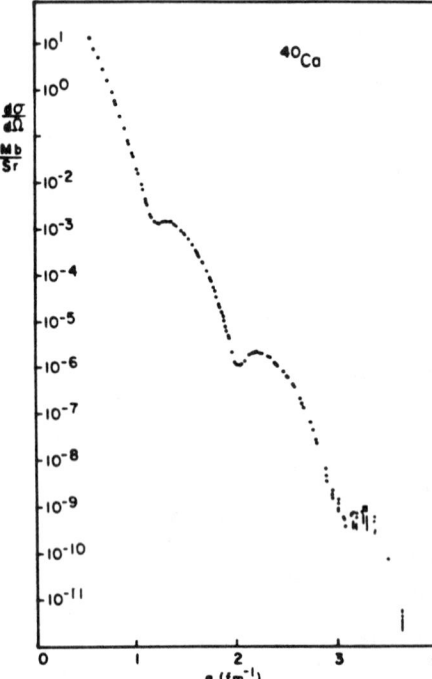

Fig. 1 Elastic(e,e) cross section for ^{40}Ca vs. momentum transfer [2]. The scattering here is from the charge distribution.

What we measure in electron scattering, basically, is a macroscopic diffraction pattern, and I want to show you this once more in Figure 1 [2]. This is the ^{40}Ca diffraction pattern plotted against the momentum transfer q. You set up these huge spectrometers, in the laboratory, and measure this optical diffraction pattern. This one has been measured over 13 decades. You then essentially take the Fourier transform of this diffraction pattern, and determine the microscopic distribution of charge in this nucleus, as shown in Figure 2; the scale here is in Fermis where $1F \equiv 10^{-13}$cm. That little band is the experimental accuracy with which we have determined the charge density in ^{40}Ca from these experiments.

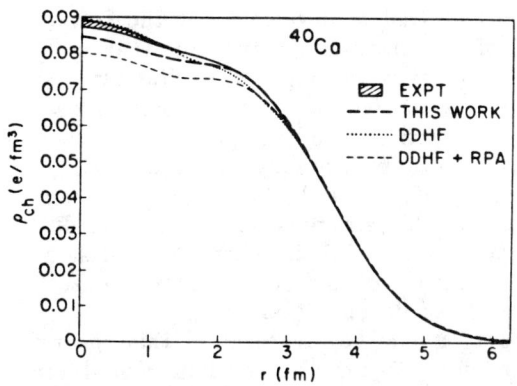

Fig. 2 Experimental charge density of ^{40}Ca with estimated uncertainty from elastic electron scattering (solid lines and shaded area) and relativistic Hartree calculations of this quantity within the framework of QHD (heavy dashed line). Taken from refs. [2, 4].

Furthermore, the electron is a versatile probe for nuclear physics. Not only is there an interaction with the charge density, but there is an interaction with the convection current, and also with the intrinsic magnetization density and corresponding magnetization current coming from the intrinsic magnetic moment of the nucleons.

Let me say a little bit about how we do nuclear physics. I will start with what I call the traditional approach [3]. In this approach you start with a static two-body potential fit to two-nucleon scattering data, you insert that into the non-relativistic many-particle Schrödinger equation, and then you solve that equation within some approximation, or in the two- and three-body problem, you can now essentially solve it exactly. You construct the nuclear currents from the properties of free nucleons, and you use these currents to probe the system. Now although this approach to nuclear physics has had many successes, as you all know, it is clearly inadequate for a more detailed understanding of the nucleus.

A more appropriate set of degrees of freedom for the nuclear system consists of the hadrons, the strongly interacting mesons and baryons. In addition, one of the current goals of nuclear physics is to study nuclear matter under extreme conditions - high temperature, high pressure, high flow velocities. These conditions are relevant to astrophysics, and relativistic heavy-ion reactions. Furthermore, we want to study the response of the nuclear system to high-q^2 probes. In order to have a theoretical framework to describe these phenomena, it is essential that we incorporate general principles of physics such as quantum mechanics, special relativity, and causality, in our theoretical description. The only consistent theoretical framework we have for describing such an interacting, relativistic, many-body system is relativistic quantum field theory based on a local lagrangian density. I like to refer to such theories of the nuclear system as quantum hadrodynamics, or QHD [4].

Certainly one of the great intellectual achievements of the last decade has been the unification of the electromagnetic and weak interactions [5-7]. It is essential to continue to put this Standard Model of the electroweak interactions to rigorous tests. Furthermore, we have a theory of the strong interactions binding quarks into the observed hadrons; that theory is quantum chromodynamics or QCD, based on an internal color symmetry [8]. I will discuss how we can use electroweak interactions to probe this structure of the Standard Model of the strong, electromagnetic, and weak interactions, and how we can use nuclei to study the structure of the strong interactions.

I want to say just a couple of words about quantum chromodynamics and remind you of a few key features of this theory [8]. The first is asymptotic freedom. Asymptotic freedom roughly says the following: when the momenta entering into a process are very large, or equivalently at very short distances, the renormalized coupling constant governing that process goes to zero. This means, under these conditions, one can do perturbation theory. The other striking aspect of the theory of QCD is that the underlying degrees of freedom are never seen as asymptotic free scattering states in the laboratory. The quarks and gluons are confined to the interior of the observed hadrons. There are strong indications from lattice gauge theory, which tries to solve QCD in the strong-coupling regime on a finite space-time lattice, that confinement is indeed a dynamical aspect of QCD.

I want to discuss one application of the relativistic aspects of nuclear physics, and Bernard Frois talked about this yesterday. I will go through it again very briefly. Let us take elastic magnetic scattering from ^{3}He, and make the world's simplest model. We will say that ^{3}He is a neutron hole in the ^{4}He core, make an oscillator model, and get the oscillator parameter from elastic charge scattering. The solid line in Figure 3 is the magnetic form factor you then predict for ^{3}He [9]. Now let us add the pion exchange current. This is the current coming from the exchange of charged pions in this nuclear system. In fact, you can calculate the long-

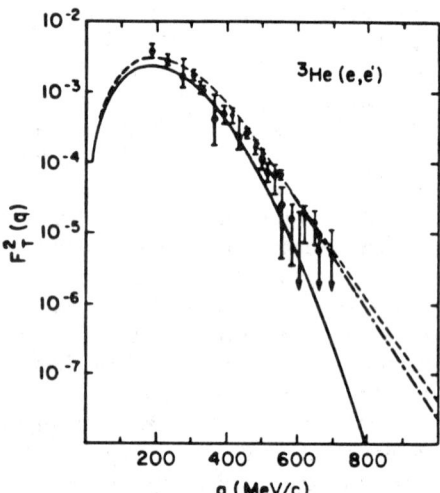

Fig. 3 Elastic transverse form factor for ^{3}He (e,e) with (dashed) and without (solid) one-pion-exchange currents [9].

range part of the pion exchange current from basic low-energy theorems. If you put in that pion exchange current in the calculation, what you get is the dashed curve in Figure 3 [9]. Well, you really are not going to conclude very much from that, right? On the other hand, if you now push and go to high momentum transfer, then you get the data shown in Figure 4 [10]. The data I just showed you is here divided by q^2, that is why the curve goes to one at $q^2=0$. The dashed curve is the best calculation we have based on the solution to the Faddeev equations for this three-body system, and on structureless nucleons. It is clear that what was a small effect before at low q^2 now becomes an order of magnitude effect at larger q^2. This is a clear demonstration of the role of the sub-nucleonic, or hadronic, degrees of freedom in the nuclear system.

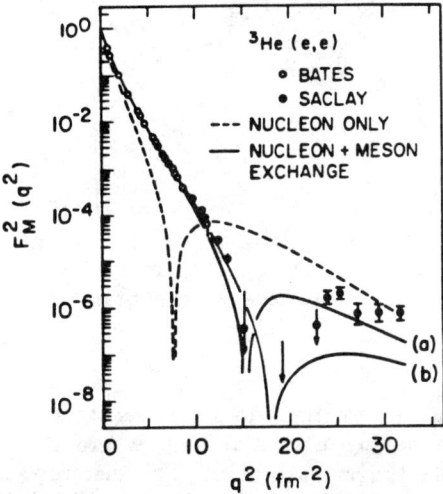

Fig. 4 Elastic magnetic form factor for ^{3}He (e,e) out to high q^2 [10]. Two exchange current theories are shown.

There are several lessons from this story. The first is that the intermediate-q^2 results illustrate the marginal role of exchange currents in the traditional nuclear physics domain. The second moral is that the high-q^2 results illustrate the need for an explicit treatment of the hadronic degrees of freedom in this system, or for QHD. We can summarize that moral in the following way: the appropriate set of degrees of freedom depends on the distance scale at which we probe the system. There is still another lesson in this. The only way we can arrive at this unambiguous identification of exchange currents, or of the role of the sub-nucleonic hadronic degrees of freedom, is to have a very accurate theoretical calculation in which we believe, which is clearly inadequate in some range of kinematics.

Where are we today in nuclear physics? We study the properties of the nuclear system at accelerators such as Bates, Saclay, NIKHEF and others, and we can accurately interpret that data in terms of nucleonic and sub-nucleonic hadronic degrees of freedom. On the other hand, we know from the deep inelastic scattering work done at SLAC that at very high energy transfer and very high momentum transfer, that is in the deep-inelastic region, we see the point-like substructure of these hadrons [11]. Therefore, in the intermediate-range of momentum transfer and energy loss, there is clearly an interesting region of physics. To emphasize this point, I will quote a sentence from the

report of the Vogt Subcommittee of NSAC which was the last
committee to re-examine the question of constructing a 4 GeV electron
accelerator for nuclear physics. It concluded that "The search for new
nuclear degrees of freedom and the relationship of nucleon-meson
degrees of freedom to quark-gluon degrees of freedom in nuclei is one
of the most challenging and fundamental questions of physics."

Let me give you, in Figure 5, a picture of what the nucleus looks
like in the Standard Model. This is a cartoon, but underneath that
cartoon there is a lagrangian and there are local currents. It
is the lagrangian of QCD and
the currents are those of the
Standard Model. What does
the nucleus look like in this
Standard Model? First, I want
to point out that the structure
of confinement in the many-
baryon system, as in the
single-baryon system, is an
unsolved problem. Confinement
presumably arises because of
the non-linear gluon couplings
in the QCD lagrangian. The
electroweak interaction, as
provided by an electron, or a
neutrino, or an e^+e^-
annihilation, couples to the
quarks; it sees through
the gluon structure.

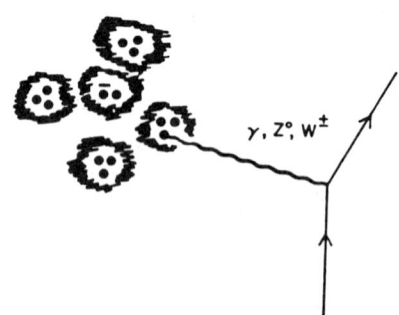

Fig. 5 Picture of the
nucleus in the Standard
Model.

The electroweak interaction does not see this
confining gluon structure. It is like having a crystal ball, where the
quarks are tiny colored objects inside the crystal ball. The electroweak
interaction sees that interior quark structure, and the gluons with their
non-linear couplings, which are responsible for the confinement, are
transparent to this electroweak interaction. Thus the electroweak
interaction does indeed directly see the quark structure of nuclei.

Now this picture and the underlying lagrangian and the
currents have some rather striking consequences. Let me just tell you
two of them [1, 12]. Suppose I use the nuclear system to select an
isospin-zero to an isospin-zero transition, that is a pure isoscalar
transition. Suppose I confine myself to the nuclear domain, where by
nuclear domain I mean that part of the Hilbert space made of only up
(u) and down (d) quarks, and any number of ū and d anti-quarks.
Within that subspace of the full Hilbert space you can prove the
following relation. The neutrino cross section is proportional with a
known constant of proportionality to the electron scattering cross
section.

$$\frac{d\sigma_{\nu_1 \to \nu_1}}{d\sigma_{\nu_1 \to \nu_1}'} = \sin^4\theta_w \frac{G^2 q^4}{2\pi^2 \alpha^2} d\sigma_{e \to e'} \quad ; T = 0 \to T = 0$$

Now what does this mean? It means that if I scatter a neutrino from ^{40}Ca, I see that entire diffraction pattern shown before in Figure 1. I can lay those two cross sections (appropriately scaled) on top of each other and they should be the same over those 13 decades! And this holds for all distance scales; it holds at a distance scale where you interpret nuclear structure in terms of gross nuclear properties, down to a distance scale where you interpret it in terms of structureless nucleons, to a distance scale where you must invoke the sub-nucleonic hadronic degrees of freedom, down to the quark-gluon level itself.

As a second example, suppose I look at a $0^+ \to 0^+$ transition, for example elastic scattering from ^{12}C. In that case, the parity-violating asymmetry which arises from an interference of Z° exchange with photon exchange is again a known factor times the ratio of two form factors.

$$\mathcal{A}_{ee'} = \frac{d\sigma_\uparrow - d\sigma_\downarrow}{d\sigma_\uparrow + d\sigma_\downarrow}$$

$$= -\frac{q^2 G}{2\pi\alpha\sqrt{2}} \left[\frac{F^{(o)}(q^2)}{F^\gamma(q^2)} \right] \quad ; 0^+ \to 0^-$$

One is the form factor for the weak neutral current and the other the ordinary electromagnetic form factor. The former measures the distribution of the weak neutral charge over this nucleus, this complicated hadronic system, and the latter the distribution of electromagnetic charge. Within the Standard Model this ratio of form factors is a constant for all q^2, and this asymmetry should be strictly linear in q^2 [13].

$$\mathcal{A}_{ee'} = \frac{q^2 G}{\pi\alpha\sqrt{2}} \sin^2\theta_w \quad ; \begin{array}{c} 0^+ \to 0^+ \\ T=0 \to T=0 \end{array}$$

This strict linear dependence, to me, is a true test of the unification of the weak and electromagnetic interactions. An experiment to measure this parity-violating asymmetry for ^{12}C is being carried out at Bates, and the fact that it is a true test of the Standard Model in the nuclear domain, where the strong interactions are strong, is the reason that I personally give it a top priority.

Let me set up a straw man who says, "We have quarks and gluons and the QCD lagrangian, therefore the problem is no longer interesting." Nobody would ever say that, right? (I have heard it!) Let me make the same statement, "We have electrons and protons (in nuclei) and the lagrangian of QED. So what?" Well, first we have atoms. Then we have crystals. Then we have semiconductors. And then we have superconductors. And then we have superfluids. (In fact, you would not know about these latter phenomena unless you had the appropriate experimental facilities to study low temperatures, for example, the facilities we heard about in Hermann's talk this morning.) And then of course, you can combine atoms into molecules and you have chemistry, then you have biology, and then you have life. It is not an uninteresting system of consequences following from that lagrangian and these underlying degrees of freedom!

This is just a schematic of where we are today.

```
NP  ↘                           ↗  Structure of Matter
       Standard Model
       (Common language)           Tests of Standard Model
HEP ↗                           ↘  Beyond the Standard Model
```

One of the things I like is that nuclear physics and high energy physics are, in a certain sense, coming together again through the Standard Model. We now have a framework in which we all operate. It is a new language for nuclear physicists, but it is absolutely essential to learn that language, because it describes the underlying theory of the structure of matter. And now, in fact, we proceed in slightly different directions. Nuclear physics is the study of the structure of the matter that underlying theory describes. High energy physics goes in the direction of probing beyond the Standard Model, as we heard from bj and in the talk yesterday. Both nuclear physics and high energy physics are interested in testing the Standard Model in all of its complexity and all of its richness.

Success generates opportunities. We have talked about 35 years of electron scattering and electron interactions. The success of this field, in studying nuclei and basic interactions, is indicated by where we are today. If you simply look at what is already operating, and what is being contemplated, it is clear that the success of this field is what generates the opportunities ahead of us. We have heard about LEP, we have heard about Mainz, we have heard about HERA, and we have heard about NIKHEF. Let me be chauvinistic and just talk for a moment about the U.S. In nuclear physics, CEBAF is now an approved construction project (as of last weekend!) Bates has an upgrade to 1 GeV with the pulse stretcher ring. Illinois has its microtron which will reach, as we heard, 450 MeV. I think all of these projects will actually go; that is my best estimate at the present time. High energy physics has SLC. As far as nuclear physics is concerned, we have not had the quality beams, the kinematic range, and the coincidence capability before in this country. There is certainly plenty of unexplored physics for everybody. CEBAF, if you like, is pushing on the kinematic frontier. It will stretch our picture of the nucleus to the extreme, test it under extreme kinematic conditions, and work towards asymtopia, or at least asymptopia as demonstrated by the SLAC deep inelastic experiments. Bates has a rich program with high resolution spectroscopy, particularly with internal targets, polarized beams, and exotic polarized jet targets. Illinois has an excellent program of other things, including studying angular correlations of collective levels; we have seen those beautiful angular correlations. This has never been possible before. Certainly from a physics point of view, there is more than a decade of good physics for everybody.

My strongest argument for Bates and Illinois, however, is that they are really the best sources we have for young people for this field, and, in fact, for all of nuclear science. Bright, creative, young people are not only essential to the science, but are also the most valuable resource this country has.

I tried to summarize my own thoughts on CEBAF, and I summarize them this way: CEBAF will provide the most precise, accessible probe of matter. The interaction is known, and one knows what is being measured. It is an interesting time for nuclear physics; we are told there is a whole new underlying set of degrees of freedom and forces in the nuclear system. What we are really building is a tool and a capability for the next generation of nuclear scientists.

The scientific goal of CEBAF is to study the structure of the nuclear many-body system, its quark substructure, and the strong and electroweak interactions that govern the behavior of this fundamental form of matter.

And I would like to close with three quotations and a story. I like these quotes, I use them all the time. The first quote is from Herb Anderson at a talk at Los Alamos. (Actually it was a question he asked Herman Feshbach after Herman's talk.)

"We have been doing nuclear physics for 50 years without quarks. Why do we need them now?"

That is actually a very profound question for nuclear physics and nuclear physicists. I ask you to think about it very carefully.

The second quote is a comment Bob Wilson made to me a couple of years ago.

"The single most important practical application of the recent advances in particle physics may well be the revolution in our picture of the nucleus."

And finally, Nathan Isgur gave a redefinition of the field for the future which I particularly like.

"Nuclear physics is the study of the strong-interaction, confinement aspects of QCD."

And now let me close with a story. Some of you may know that I have been pushing a high resolution capability for CEBAF. Two years ago there was a symposium at Stanford to honor Bob Hofstadter's 70th birthday. It was a two-day symposium; I gave a talk there and told this story. Twenty years ago, in about 1968, Bob and I collaborated on the theoretical justification for the proposal of a 2 GeV CW electron accelerator. During that collaboration Bob said that we had to have a high-resolution capability. He said that if you look with high-resolution at high energy, you will certainly discover new phenomena. And I said to myself, "That is nonsense. You are not going to see anything at high energy with high resolution." And now you have to remember the time. In those days high energy physics looked asymptotic. Everything you were doing at that time was smooth, you were already in asymptopia nothing peculiar was going on. It was actually very dull! And then one night at SPEAR they had a broad bump in their spectrum, and it would have remained just a broad bump forever, but they had the capability of turning a knob and getting the resolution up. And so they sat there that night and kept turning the knob and the resolution kept improving and, of course the ψ was what they saw. They saw the role of a new underlying set of degrees of freedom; they saw it as a very sharp resonance at an energy that nobody had expected. One had never seen such a sharp line before at that energy, and in a certain sense it revolutionized high energy physics. My moral from that story is that I will continue to rely on Bob Hofstadter's intuition.

Thank you for this symposium.

References

[1] J. D. Walecka, "Electron Scattering," ANL-83-50, Argonne Nat. Lab. (1984)

[2] B. Frois et al., Lect. Notes in Phys. $\underline{108}$, Springer (1979)

[3] A. L. Fetter and J. D. Walecka, Quantum Theory of Many-Particle Systems, McGraw-Hill, New York (1971)

[4] B. D. Serot and J. D. Walecka, "The Relativistic Nuclear Many-Body Problem," Adv. in Nucl. Phys., Vol. $\underline{16}$, eds. J. W. Negele and E. Vogt, Plenum Press, New York (1986)

[5] S. Weinberg, Phys. Rev. Lett. $\underline{19}$, 1264 (1967); Phys Rev. $\underline{D5}$, 1412 (1972)

[6] A. Salam and J. C. Ward, Phys. Lett. $\underline{13}$, 168 (1964)

[7] S. L. Glashow et al., Phys. Rev. $\underline{D2}$, 1285 (1970)

[8] F. Wilczek, Ann. Rev. of Nucl. Sci. $\underline{32}$, 177 (1982)

[9] J. Dubach et al., Nucl. Phys. $\underline{A271}$, 279 (1976)

[10] J. M. Cavedon et al., Phys. Rev. Lett. $\underline{49}$, 986 (1982)

[11] J. I. Friedman and H. W. Kendall, Ann. Rev. Nucl. Sci. $\underline{22}$, 203 (1972)

[12] J. D. Walecka, A.I.P. Conf. Proc. $\underline{123}$, A.I.P., New York (1984)

[13] G. Feinberg, Phys. Rev. $\underline{D12}$, 3575 (1975)

LIST OF PARTICIPANTS

Alarcon, Dr. Ricardo
University of Illinois
Nuclear Physics Laboratory
23 Stadium Drive
Champaign, IL 61820

Ammons, Mr. Edsel
University of Illinois
Nuclear Physics Laboratory
23 Stadium Drive
Champaign, IL 61820

Barnes, Prof. Peter
Carnegie-Mellon University
Department of Physics
Pittsburgh, PA 15213

Baym, Gordon
University of Illinois
Loomis Laboratory of Physics
1110 West Green Street
Urbana, IL 61801

Berdichevsky, Daniel
McMaster University
Physics Department, SS-348
Hamilton, Ontario L8S 4M1
CANADA

Birenbaum, Yair
University of Illinois
Nuclear Physics Laboratory
23 Stadium Drive
Champaign, IL 61820

Bjorken, Dr. James
Fermi National Accelerator Lab.
Research Division
P. O. Box 500
Batavia, IL 60510

Bosco, Bernardino
University of Florence
Ist. di Fisica
Teorica-Lgo E. Fermi 2
Firenze, Italy

Branson, Prof. James G.
Massachusetts Inst. of Technology
Department of Physics
77 Massachusetts Avenue
Cambridge, MA 02139

Breton, Vincent
DPhN/HE
CEN Saclay
91191 Gif-sur-Yvette Cedex
FRANCE

Brussel, Dr. M. K.
DPhN/HE
CEN Saclay
91191 Gif-sur-Yvette Cedex
FRANCE

Calarco, Prof. J. R.
University of New Hampshire
Department of Physics
Durham, NH 03824

Caplan, Henry S.
University of Saskatchewan
Department of Physics
Saskatoon S7N 0W0
CANADA

Cardman, L. S.
University of Illinois
Nuclear Physics Laboratory
23 Stadium Drive
Champaign, IL 61820

Chung, Ping Lin
University of Iowa
Department of Physics
Iowa City, Iowa 52240

Clark, Dr. Bunny
Ohio State University
Physics Department
Columbus, OH 43210

Co, Giampaolo
University of Illinois
Loomis Laboratory of Physics
1110 West Green Street
Urbana, IL 61801

Coester, Fritz
Argonne National Laboratory
Physics Division
9700 South Cass Avenue
Argonne, IL 60439

Cohen, Joesph
Indiana University
Physics Department
Swain Hall-West
Bloomington, IN 47405

Cole, Mr. Patrick
University of Illinois
Nuclear Physics Laboratory
23 Stadium Drive
Champaign, IL 61820

Crannell, Hall
Catholic University of America
Physics Department
Washington, DC 20064

Dale, Dan
University of Illinois
Nuclear Physics Laboratory
23 Stadium Drive
Champaign, IL 61820

DeVries, C.
NIKHEF-K
P.O. Nox 4395
1009 AJ Amsterdam
The Netherlands

Debevec, Prof. Paul T.
University of Illinois
Nuclear Physics Laboratory
23 Stadium Drive
Champaign, IL 61820

Deininger, Robert
University of Illinois
Nuclear Physics Laboratory
23 Stadium Drive
Champaign, IL 61820

Dolfini, Mr. Steve
University of Illinois
Nuclear Physics Laboratory
23 Stadium Drive
Champaign, IL 61820

Eisenstein, Professor R. A.
University of Illinois
Nuclear Physics Laboratory
23 Stadium Drive
Champaign, IL 61820

Errede, Steven
University of Illinois
Loomis Laboratory of Physics
1110 West Green Street
Urbana, IL 61801

Fabrocini, Adelchi
University of Illinois
Loomis Laboratory of Physics
1110 West Green Street
Urbana, IL 61801

Federspiel, Fred
University of Illinois
Nuclear Physics Laboratory
23 Stadium Drive
Champaign, IL 61820

Feshbach, Professor Herman
Massachusetts Inst. of Technology
Department of Physics
Cambridge, MA 02139

Finn, John M.
College of William & Mary
Department of Physics
Williamsburg, VA 23185

Franklin, Gregg
University of Illinois
Nuclear Physics Laboratory
23 Stadium Drive
Champaign, IL 61820

Frois, Dr. Bernard
DPhN/HE
CEN Saclay
91191 Gif-sur-Yvette Cedex
FRANCE

Furnstahl, Richard
Indiana University
Department of Physics
Bloomington, IN 47405

Gavilano, Jorge
University of Illinois
Loomis Laboratory of Physics
1110 West Green Street
Urbana, IL 61801

Giovanetti, Kevin
University of Virginia
Physics Department
Charlottesville, VA 22901

Goulard, Bernard
Universite de Montreal
Laboratoire de Physique Nucleaire
Montreal, PQ, CANADA H3A 2T8

Grunder, Dr. Hermann A.
CEBAF
12070 Jefferson Avenue
Newport News, VA 23606

Hanson, Prof. A.O.
University of Illinois
Nuclear Physics Laboratory
23 Stadium Drive
Champaign, IL 61820

Heisenberg, Prof. Jochen
University of New Hampshire
Department of Physics
DeMeritt Hall
Dunham, NC 03824

Hertzog, Prof. David
University of Illinois
Nuclear Physics Laboratory
23 Stadium Drive
Champaign, IL 61820

Hoblit, Mr. Sam
University of Illinois
Nuclear Physics Laboratory
23 Stadium Drive
Champaign, IL 61820

Hofstadter, Prof. Robert
Stanford University
Department of Physics
Stanford, CA 94305

Holt, Dr. Roy
Argonne National Laboratory
Physics Division
9700 South Cass Avenue
Argonne, IL 60439

Hughes, Mr. Jim
1409 Cottonwood Lane
Apartment A
Mt. Prospect, IL 60056

Isabelle, Didier
CERI-CNRS and ALS Saclay
3A rue de la Ferollerie
45071 ORLEANS-CEDEX 2
France

Jenkins, Dr. David A.
Virginia Techn
Physics Department
Blacksburg, VA 24061

Jones, Mr. Rick
University of Illinois
Nuclear Physics Laboratory
23 Stadium Drive
Champaign, IL 61820

Kerst, Prof. Donald W.
University of Wisconsin
Department of Physics
1150 University Avenue
Madison, WI 53706

Knott, Mr. Jonathan
University of Illinois
Nuclear Physics Laboratory
23 Stadium Drive
Champaign, IL 61820

Koch, Dr. William
American Institute of Physics
335 East 45th Street
New York, NY 10017

Kolb, Mr. Norman
University of Illinois
Nuclear Physics Laboratory
23 Stadium Drive
Champaign, IL 61820

Kumano, Shunzo
University of Illinois
Loomis Laboratory of Physics
1110 West Green Street
Urbana, IL 61801

Kuyatt, Chris
National Bureau of Standards
Bldg. 245, Rm. C229
Gaithersburg, MD 20899

Lanzl, Lawrence
Rush-Presbyterian-St. Luke's
Medical Center
1753 W. Congress Parkway
Chicago, IL 60612

Laszewski, Ron
University of Illinois
Nuclear Physics Laboratory
23 Stadium Drive
Champaign, IL 61820

Leiss, Dr. James E.
13013 Chestnut Oak Drive
Gaithersburg, MD 20878

Lewart, Daniel
University of Illinois
Loomis Laboratory of Physics
1110 West Green Street
Urbana, IL 61801

Lightbody, J. W.
National Bureau of Standards
Gaithersburg, MD 20899

Linzey, Andrew
University of Illinois
Nuclear Physics Laboratory
23 Stadium Drive
Champaign, IL 61820

Londergan, Timothy
Indiana Univ. Cyclotron Facility
2401 Milo Sampson Lane
Bloomington, IN 47405

Lyman, Ernest
1009 S. Orchard
Urbana, IL 61801

Mandeville, Mr. Joe
University of Illinois
Nuclear Physics Laboratory
23 Stadium Drive
Champaign, IL 61820

Maximon, Dr. Leonard C.
National Bureau of Standards
Gaithersburg, MD 20899

McDaniel, Prof. Boyce D.
Cornell University
Department of Physics
109 Clark Hall
Ithaca, NY 14853

Mills, Fred
Fermi National Accelerator Lab
Accelerator Division
Box 500
Batavia, IL 60510

Moniz, Ernest
MIT-Bates Linear Accel. Center
P. O. Box 846
Middleton, MA 01949

Nathan, Dr. Alan M.
University of Illinois
Nuclear Physics Laboratory
23 Stadium Drive
Champaign, IL 61820

Nishimura, Dr. M.
Osaka University
Res. Center for Nuclear Physics
Theory Group
Yamadaoka, Suita
Osaka 565, JAPAN

O'Brien, Dr. James T.
Montgomery College
Physics Department
Rockville, MD 20850

Pandharipande, Prof. V. R.
University of Illinois
Loomis Laboratory of Physics
1110 West Green Street
Urbana, IL 61801

Papanicolas, Costas
University of Illinois
Nuclear Physics Laboratory
23 Stadium Drive
Champaign, IL 61820

Pieper, Steven
Argonne National Laboratory
Physics Department
9700 South Cass Avenue
Argonne, IL 60439

Pitts, Karl
University of Illinois
Nuclear Physics Laboratory
23 Stadium Drive
Champaign, Il 61820

Ravenhall, Professor D. G.
University of Illinois
Loomis Laboratory of Physics
1110 West Green Street
Urbana, IL 61801

Ritchie, Buddy
University of Illinois
Nuclear Physics Laboratory
23 Stadium Drive
Champaign, IL 61820

Ruuskanen, Vesa
University of Illinois
Loomis Laboratory of Physics
1110 West Green
Urbana, IL 61801

Saito, Teijiro
Massachusetts Inst. of Technology
Department of Physics
Middleton, MA 01949

Sard, Dr. Robert
University of Illinois
Loomis Laboratory of Physics
1110 West Green Street
Urbana, IL 61801

Schiavilla, Rocco
University of Illinois
Loomis Laboratory of Physics
1110 West Green Street
Urbana, IL 61801

Schlagel, Tom
University of Illinois
Loomis Laboratory of Physics
1110 West Green Street
Urbana, IL 61801

Scott, Dr. M. B.
4369 North Wilson
Fresno, CA 93704

Sellyey, William
University of Illinois
Nuclear Physics Laboratory
23 Stadium Drive
Champaign, IL 61820

Serdarevic, Amra
University of Illinois
Nuclear Physics Laboratory
23 Stadium Drive
Champaign, IL 61820

Sustich, Andrew
University of Illinois
Loomis Laboratory of Physics
1110 West Green Street
Urbana, IL 61801

Seth, Kamal
Northwestern University
Dept. of Physics and Astronomy
Evanston, IL 60201

Tabakin, Frank
University of Pitttsburgh
Department of Physics
Pittsburgh, PA 15260

Sick, Dr. Ingo
University of Basel
Department of Physics
Basel, Switzerland

Templon, Jeff
Indiana Univ. Cyclotron Facility
2401 Milo B. Sampson Lane
Bloomington, IN 47405

Simmons, Ralph O.
University of Illinois
Loomis Laboratory of Physics
1110 West Green Street
Urbana, IL 61801

Uberall, Herbert
Catholic University of America
Physics Department
Washington, DC 20064

Skaggs, Lester S.
359 Osage Street
Park Forest, IL 60466

VanDijk, W.
McMaster University
Physics Department, SS-348
Hamilton, Ontario L8S 4L8
CANADA

Smith, Prof. James H.
University of Illinois
Loomis Laboratory of Physics
1110 West Green Street
Urbana, IL 61801

Walecka, Prof. J. Dirk
CEBAF
12070 Jefferson Avenue
Newport News, VA 23606

Sullivan, Dr. Jeremiah D.
University of Illinois
Loomis Laboratory of Physics
1110 West Green Street
Urbana, IL 61801

Wambach, Jochen
University of Illinois
Loomis Laboratory of Physics
1110 West Green Street
Urbana, IL 61801

Wattenberg, Prof. Al
University of Illinois
Loomis Laboratory of Physics
1110 West Green Street
Urbana, IL 61801

Wells, Mr. Doug
University of Illinois
Nuclear Physics Laboratory
23 Stadium Drive
Champaign, IL 61820

Williamson, Dr. Steven
University of Illinois
Nuclear Physics Laboratory
23 Stadium Drive
Champaign, IL 61820

Wiringa, Robert
Argonne National Laboratory
Physics Division
9700 South Cass Avenue
Argonne, IL 60439

Yennie, Donald R.
Cornell University
Department of Physics
109 Clark Hall
Ithaca, NY 14853

Zimmerman, Dr. Peter D.
11 DuPont Circle, N.W.
Washington, DC 20036

AIP Conference Proceedings

		L.C. Number	ISBN
No. 1	Feedback and Dynamic Control of Plasmas – 1970	70-141596	0-88318-100-2
No. 2	Particles and Fields – 1971 (Rochester)	71-184662	0-88318-101-0
No. 3	Thermal Expansion – 1971 (Corning)	72-76970	0-88318-102-9
No. 4	Superconductivity in d- and f-Band Metals (Rochester, 1971)	74-18879	0-88318-103-7
No. 5	Magnetism and Magnetic Materials – 1971 (2 parts) (Chicago)	59-2468	0-88318-104-5
No. 6	Particle Physics (Irvine, 1971)	72-81239	0-88318-105-3
No. 7	Exploring the History of Nuclear Physics – 1972	72-81883	0-88318-106-1
No. 8	Experimental Meson Spectroscopy –1972	72-88226	0-88318-107-X
No. 9	Cyclotrons – 1972 (Vancouver)	72-92798	0-88318-108-8
No. 10	Magnetism and Magnetic Materials – 1972	72-623469	0-88318-109-6
No. 11	Transport Phenomena – 1973 (Brown University Conference)	73-80682	0-88318-110-X
No. 12	Experiments on High Energy Particle Collisions – 1973 (Vanderbilt Conference)	73-81705	0-88318-111-8
No. 13	π-π Scattering – 1973 (Tallahassee Conference)	73-81704	0-88318-112-6
No. 14	Particles and Fields – 1973 (APS/DPF Berkeley)	73-91923	0-88318-113-4
No. 15	High Energy Collisions – 1973 (Stony Brook)	73-92324	0-88318-114-2
No. 16	Causality and Physical Theories (Wayne State University, 1973)	73-93420	0-88318-115-0
No. 17	Thermal Expansion – 1973 (Lake of the Ozarks)	73-94415	0-88318-116-9
No. 18	Magnetism and Magnetic Materials – 1973 (2 parts) (Boston)	59-2468	0-88318-117-7
No. 19	Physics and the Energy Problem – 1974 (APS Chicago)	73-94416	0-88318-118-5
No. 20	Tetrahedrally Bonded Amorphous Semiconductors (Yorktown Heights, 1974)	74-80145	0-88318-119-3
No. 21	Experimental Meson Spectroscopy – 1974 (Boston)	74-82628	0-88318-120-7
No. 22	Neutrinos – 1974 (Philadelphia)	74-82413	0-88318-121-5
No. 23	Particles and Fields – 1974 (APS/DPF Williamsburg)	74-27575	0-88318-122-3
No. 24	Magnetism and Magnetic Materials – 1974 (20th Annual Conference, San Francisco)	75-2647	0-88318-123-1
No. 25	Efficient Use of Energy (The APS Studies on the Technical Aspects of the More Efficient Use of Energy)	75-18227	0-88318-124-X

No. 26	High-Energy Physics and Nuclear Structure – 1975 (Santa Fe and Los Alamos)	75-26411	0-88318-125-8
No. 27	Topics in Statistical Mechanics and Biophysics: A Memorial to Julius L. Jackson (Wayne State University, 1975)	75-36309	0-88318-126-6
No. 28	Physics and Our World: A Symposium in Honor of Victor F. Weisskopf (M.I.T., 1974)	76-7207	0-88318-127-4
No. 29	Magnetism and Magnetic Materials – 1975 (21st Annual Conference, Philadelphia)	76-10931	0-88318-128-2
No. 30	Particle Searches and Discoveries – 1976 (Vanderbilt Conference)	76-19949	0-88318-129-0
No. 31	Structure and Excitations of Amorphous Solids (Williamsburg, VA, 1976)	76-22279	0-88318-130-4
No. 32	Materials Technology – 1976 (APS New York Meeting)	76-27967	0-88318-131-2
No. 33	Meson-Nuclear Physics – 1976 (Carnegie-Mellon Conference)	76-26811	0-88318-132-0
No. 34	Magnetism and Magnetic Materials – 1976 (Joint MMM-Intermag Conference, Pittsburgh)	76-47106	0-88318-133-9
No. 35	High Energy Physics with Polarized Beams and Targets (Argonne, 1976)	76-50181	0-88318-134-7
No. 36	Momentum Wave Functions – 1976 (Indiana University)	77-82145	0-88318-135-5
No. 37	Weak Interaction Physics – 1977 (Indiana University)	77-83344	0-88318-136-3
No. 38	Workshop on New Directions in Mossbauer Spectroscopy (Argonne, 1977)	77-90635	0-88318-137-1
No. 39	Physics Careers, Employment and Education (Penn State, 1977)	77-94053	0-88318-138-X
No. 40	Electrical Transport and Optical Properties of Inhomogeneous Media (Ohio State University, 1977)	78-54319	0-88318-139-8
No. 41	Nucleon-Nucleon Interactions – 1977 (Vancouver)	78-54249	0-88318-140-1
No. 42	Higher Energy Polarized Proton Beams (Ann Arbor, 1977)	78-55682	0-88318-141-X
No. 43	Particles and Fields – 1977 (APS/DPF, Argonne)	78-55683	0-88318-142-8
No. 44	Future Trends in Superconductive Electronics (Charlottesville, 1978)	77-9240	0-88318-143-6
No. 45	New Results in High Energy Physics – 1978 (Vanderbilt Conference)	78-67196	0-88318-144-4
No. 46	Topics in Nonlinear Dynamics (La Jolla Institute)	78-57870	0-88318-145-2
No. 47	Clustering Aspects of Nuclear Structure and Nuclear Reactions (Winnepeg, 1978)	78-64942	0-88318-146-0
No. 48	Current Trends in the Theory of Fields (Tallahassee, 1978)	78-72948	0-88318-147-9

No. 49	Cosmic Rays and Particle Physics – 1978 (Bartol Conference)	79-50489	0-88318-148-7
No. 50	Laser-Solid Interactions and Laser Processing – 1978 (Boston)	79-51564	0-88318-149-5
No. 51	High Energy Physics with Polarized Beams and Polarized Targets (Argonne, 1978)	79-64565	0-88318-150-9
No. 52	Long-Distance Neutrino Detection – 1978 (C.L. Cowan Memorial Symposium)	79-52078	0-88318-151-7
No. 53	Modulated Structures – 1979 (Kailua Kona, Hawaii)	79-53846	0-88318-152-5
No. 54	Meson-Nuclear Physics – 1979 (Houston)	79-53978	0-88318-153-3
No. 55	Quantum Chromodynamics (La Jolla, 1978)	79-54969	0-88318-154-1
No. 56	Particle Acceleration Mechanisms in Astrophysics (La Jolla, 1979)	79-55844	0-88318-155-X
No. 57	Nonlinear Dynamics and the Beam-Beam Interaction (Brookhaven, 1979)	79-57341	0-88318-156-8
No. 58	Inhomogeneous Superconductors – 1979 (Berkeley Springs, W.V.)	79-57620	0-88318-157-6
No. 59	Particles and Fields – 1979 (APS/DPF Montreal)	80-66631	0-88318-158-4
No. 60	History of the ZGS (Argonne, 1979)	80-67694	0-88318-159-2
No. 61	Aspects of the Kinetics and Dynamics of Surface Reactions (La Jolla Institute, 1979)	80-68004	0-88318-160-6
No. 62	High Energy e^+e^- Interactions (Vanderbilt, 1980)	80-53377	0-88318-161-4
No. 63	Supernovae Spectra (La Jolla, 1980)	80-70019	0-88318-162-2
No. 64	Laboratory EXAFS Facilities – 1980 (Univ. of Washington)	80-70579	0-88318-163-0
No. 65	Optics in Four Dimensions – 1980 (ICO, Ensenada)	80-70771	0-88318-164-9
No. 66	Physics in the Automotive Industry – 1980 (APS/AAPT Topical Conference)	80-70987	0-88318-165-7
No. 67	Experimental Meson Spectroscopy – 1980 (Sixth International Conference, Brookhaven)	80-71123	0-88318-166-5
No. 68	High Energy Physics – 1980 (XX International Conference, Madison)	81-65032	0-88318-167-3
No. 69	Polarization Phenomena in Nuclear Physics – 1980 (Fifth International Symposium, Santa Fe)	81-65107	0-88318-168-1
No. 70	Chemistry and Physics of Coal Utilization – 1980 (APS, Morgantown)	81-65106	0-88318-169-X
No. 71	Group Theory and its Applications in Physics – 1980 (Latin American School of Physics, Mexico City)	81-66132	0-88318-170-3
No. 72	Weak Interactions as a Probe of Unification (Virginia Polytechnic Institute – 1980)	81-67184	0-88318-171-1
No. 73	Tetrahedrally Bonded Amorphous Semiconductors (Carefree, Arizona, 1981)	81-67419	0-88318-172-X

No. 74	Perturbative Quantum Chromodynamics (Tallahassee, 1981)	81-70372	0-88318-173-8
No. 75	Low Energy X-Ray Diagnostics – 1981 (Monterey)	81-69841	0-88318-174-6
No. 76	Nonlinear Properties of Internal Waves (La Jolla Institute, 1981)	81-71062	0-88318-175-4
No. 77	Gamma Ray Transients and Related Astrophysical Phenomena (La Jolla Institute, 1981)	81-71543	0-88318-176-2
No. 78	Shock Waves in Condensed Matter – 1981 (Menlo Park)	82-70014	0-88318-177-0
No. 79	Pion Production and Absorption in Nuclei – 1981 (Indiana University Cyclotron Facility)	82-70678	0-88318-178-9
No. 80	Polarized Proton Ion Sources (Ann Arbor, 1981)	82-71025	0-88318-179-7
No. 81	Particles and Fields –1981: Testing the Standard Model (APS/DPF, Santa Cruz)	82-71156	0-88318-180-0
No. 82	Interpretation of Climate and Photochemical Models, Ozone and Temperature Measurements (La Jolla Institute, 1981)	82-71345	0-88318-181-9
No. 83	The Galactic Center (Cal. Inst. of Tech., 1982)	82-71635	0-88318-182-7
No. 84	Physics in the Steel Industry (APS/AISI, Lehigh University, 1981)	82-72033	0-88318-183-5
No. 85	Proton-Antiproton Collider Physics –1981 (Madison, Wisconsin)	82-72141	0-88318-184-3
No. 86	Momentum Wave Functions – 1982 (Adelaide, Australia)	82-72375	0-88318-185-1
No. 87	Physics of High Energy Particle Accelerators (Fermilab Summer School, 1981)	82-72421	0-88318-186-X
No. 88	Mathematical Methods in Hydrodynamics and Integrability in Dynamical Systems (La Jolla Institute, 1981)	82-72462	0-88318-187-8
No. 89	Neutron Scattering – 1981 (Argonne National Laboratory)	82-73094	0-88318-188-6
No. 90	Laser Techniques for Extreme Ultraviolt Spectroscopy (Boulder, 1982)	82-73205	0-88318-189-4
No. 91	Laser Acceleration of Particles (Los Alamos, 1982)	82-73361	0-88318-190-8
No. 92	The State of Particle Accelerators and High Energy Physics (Fermilab, 1981)	82-73861	0-88318-191-6
No. 93	Novel Results in Particle Physics (Vanderbilt, 1982)	82-73954	0-88318-192-4
No. 94	X-Ray and Atomic Inner-Shell Physics – 1982 (International Conference, U. of Oregon)	82-74075	0-88318-193-2
No. 95	High Energy Spin Physics – 1982 (Brookhaven National Laboratory)	83-70154	0-88318-194-0
No. 96	Science Underground (Los Alamos, 1982)	83-70377	0-88318-195-9

No. 97	The Interaction Between Medium Energy Nucleons in Nuclei – 1982 (Indiana University)	83-70649	0-88318-196-7
No. 98	Particles and Fields – 1982 (APS/DPF University of Maryland)	83-70807	0-88318-197-5
No. 99	Neutrino Mass and Gauge Structure of Weak Interactions (Telemark, 1982)	83-71072	0-88318-198-3
No. 100	Excimer Lasers – 1983 (OSA, Lake Tahoe, Nevada)	83-71437	0-88318-199-1
No. 101	Positron-Electron Pairs in Astrophysics (Goddard Space Flight Center, 1983)	83-71926	0-88318-200-9
No. 102	Intense Medium Energy Sources of Strangeness (UC-Sant Cruz, 1983)	83-72261	0-88318-201-7
No. 103	Quantum Fluids and Solids – 1983 (Sanibel Island, Florida)	83-72440	0-88318-202-5
No. 104	Physics, Technology and the Nuclear Arms Race (APS Baltimore –1983)	83-72533	0-88318-203-3
No. 105	Physics of High Energy Particle Accelerators (SLAC Summer School, 1982)	83-72986	0-88318-304-8
No. 106	Predictability of Fluid Motions (La Jolla Institute, 1983)	83-73641	0-88318-305-6
No. 107	Physics and Chemistry of Porous Media (Schlumberger-Doll Research, 1983)	83-73640	0-88318-306-4
No. 108	The Time Projection Chamber (TRIUMF, Vancouver, 1983)	83-83445	0-88318-307-2
No. 109	Random Walks and Their Applications in the Physical and Biological Sciences (NBS/La Jolla Institute, 1982)	84-70208	0-88318-308-0
No. 110	Hadron Substructure in Nuclear Physics (Indiana University, 1983)	84-70165	0-88318-309-9
No. 111	Production and Neutralization of Negative Ions and Beams (3rd Int'l Symposium, Brookhaven, 1983)	84-70379	0-88318-310-2
No. 112	Particles and Fields – 1983 (APS/DPF, Blacksburg, VA)	84-70378	0-88318-311-0
No. 113	Experimental Meson Spectroscopy – 1983 (Seventh International Conference, Brookhaven)	84-70910	0-88318-312-9
No. 114	Low Energy Tests of Conservation Laws in Particle Physics (Blacksburg, VA, 1983)	84-71157	0-88318-313-7
No. 115	High Energy Transients in Astrophysics (Santa Cruz, CA, 1983)	84-71205	0-88318-314-5
No. 116	Problems in Unification and Supergravity (La Jolla Institute, 1983)	84-71246	0-88318-315-3
No. 117	Polarized Proton Ion Sources (TRIUMF, Vancouver, 1983)	84-71235	0-88318-316-1

No. 118	Free Electron Generation of Extreme Ultraviolet Coherent Radiation (Brookhaven/OSA, 1983)	84-71539	0-88318-317-X
No. 119	Laser Techniques in the Extreme Ultraviolet (OSA, Boulder, Colorado, 1984)	84-72128	0-88318-318-8
No. 120	Optical Effects in Amorphous Semiconductors (Snowbird, Utah, 1984)	84-72419	0-88318-319-6
No. 121	High Energy e^+e^- Interactions (Vanderbilt, 1984)	84-72632	0-88318-320-X
No. 122	The Physics of VLSI (Xerox, Palo Alto, 1984)	84-72729	0-88318-321-8
No. 123	Intersections Between Particle and Nuclear Physics (Steamboat Springs, 1984)	84-72790	0-88318-322-6
No. 124	Neutron-Nucleus Collisions – A Probe of Nuclear Structure (Burr Oak State Park - 1984)	84-73216	0-88318-323-4
No. 125	Capture Gamma-Ray Spectroscopy and Related Topics – 1984 (Internat. Symposium, Knoxville)	84-73303	0-88318-324-2
No. 126	Solar Neutrinos and Neutrino Astronomy (Homestake, 1984)	84-63143	0-88318-325-0
No. 127	Physics of High Energy Particle Accelerators (BNL/SUNY Summer School, 1983)	85-70057	0-88318-326-9
No. 128	Nuclear Physics with Stored, Cooled Beams (McCormick's Creek State Park, Indiana, 1984)	85-71167	0-88318-327-7
No. 129	Radiofrequency Plasma Heating (Sixth Topical Conference, Callaway Gardens, GA, 1985)	85-48027	0-88318-328-5
No. 130	Laser Acceleration of Particles (Malibu, California, 1985)	85-48028	0-88318-329-3
No. 131	Workshop on Polarized ^{3}He Beams and Targets (Princeton, New Jersey, 1984)	85-48026	0-88318-330-7
No. 132	Hadron Spectroscopy–1985 (International Conference, Univ. of Maryland)	85-72537	0-88318-331-5
No. 133	Hadronic Probes and Nuclear Interactions (Arizona State University, 1985)	85-72638	0-88318-332-3
No. 134	The State of High Energy Physics (BNL/SUNY Summer School, 1983)	85-73170	0-88318-333-1
No. 135	Energy Sources: Conservation and Renewables (APS, Washington, DC, 1985)	85-73019	0-88318-334-X
No. 136	Atomic Theory Workshop on Relativistic and QED Effects in Heavy Atoms	85-73790	0-88318-335-8
No. 137	Polymer-Flow Interaction (La Jolla Institute, 1985)	85-73915	0-88318-336-6
No. 138	Frontiers in Electronic Materials and Processing (Houston, TX, 1985)	86-70108	0-88318-337-4
No. 139	High-Current, High-Brightness, and High-Duty Factor Ion Injectors (La Jolla Institute, 1985)	86-70245	0-88318-338-2

No. 140	Boron-Rich Solids (Albuquerque, NM, 1985)	86-70246	0-88318-339-0
No. 141	Gamma-Ray Bursts (Stanford, CA, 1984)	86-70761	0-88318-340-4
No. 142	Nuclear Structure at High Spin, Excitation, and Momentum Transfer (Indiana University, 1985)	86-70837	0-88318-341-2
No. 143	Mexican School of Particles and Fields (Oaxtepec, México, 1984)	86-81187	0-88318-342-0
No. 144	Magnetospheric Phenomena in Astrophysics (Los Alamos, 1984)	86-71149	0-88318-343-9
No. 145	Polarized Beams at SSC & Polarized Antiprotons (Ann Arbor, MI & Bodega Bay, CA, 1985)	86-71343	0-88318-344-7
No. 146	Advances in Laser Science–I (Dallas, TX, 1985)	86-71536	0-88318-345-5
No. 147	Short Wavelength Coherent Radiation: Generation and Applications (Monterey, CA, 1986)	86-71674	0-88318-346-3
No. 148	Space Colonization: Technology and The Liberal Arts (Geneva, NY, 1985)	86-71675	0-88318-347-1
No. 149	Physics and Chemistry of Protective Coatings (Universal City, CA, 1985)	86-72019	0-88318-348-X
No. 150	Intersections Between Particle and Nuclear Physics (Lake Louise, Canada, 1986)	86-72018	0-88318-349-8
No. 151	Neural Networks for Computing (Snowbird, UT, 1986)	86-72481	0-88318-351-X
No. 152	Heavy Ion Inertial Fusion (Washington, DC, 1986)	86-73185	0-88318-352-8
No. 153	Physics of Particle Accelerators (SLAC Summer School, 1985) (Fermilab Summer School, 1984)	87-70103	0-88318-353-6
No. 154	Physics and Chemistry of Porous Media—II (Ridge Field, CT, 1986)	83-73640	0-88318-354-4
No. 155	The Galactic Center: Proceedings of the Symposium Honoring C. H. Townes (Berkeley, CA, 1986)	86-73186	0-88318-355-2
No. 156	Advanced Accelerator Concepts (Madison, WI, 1986)	87-70635	0-88318-358-0
No. 157	Stability of Amorphous Silicon Alloy Materials and Devices (Palo Alto, CA, 1987)	87-70990	0-88318-359-9
No. 158	Production and Neutralization of Negative Ions and Beams (Brookhaven, NY, 1986)	87-71695	0-88318-358-7

No. 159	Applications of Radio-Frequency Power to Plasma: Seventh Topical Conference (Kissimmee, FL, 1987)	87-71812	0-88318-359-5
No. 160	Advances in Laser Science–II (Seattle, WA, 1986)	87-71962	0-88318-360-9